Standard Grade | Credit

Chemistry

Credit Level 2004

Credit Level 2005

Credit Level 2006

Credit Level 2007

Credit Level 2008

Leckie x Leckie

© Scottish Qualifications Authority

First exam published in 2004.
Published by Leckie & Leckie Ltd, 3rd Floor, 4 Queen Street, Edinburgh EH2 1JE
tel: 0131 220 6831 fax: 0131 225 9987 enquiries@leckieandleckie.co.uk www.leckieandleckie.co.uk

ISBN 978-1-84372-624-1

A CIP Catalogue record for this book is available from the British Library.

Leckie & Leckie is a division of Huveaux plc.

Leckie & Leckie is grateful to the copyright holders, as credited at the back of the book, for permission to use their material.
Every effort has been made to trace the copyright holders and to obtain their permission for the use of copyright material.
Leckie & Leckie will gladly receive information enabling them to rectify any error or omission in subsequent editions.

[BLANK PAGE]

FOR OFFICIAL USE

C

	KU	PS
Total Marks		

0500/402

NATIONAL
QUALIFICATIONS
2004

MONDAY, 10 MAY
10.50 AM – 12.20 PM

CHEMISTRY
STANDARD GRADE
Credit Level

Fill in these boxes and read what is printed below.

Full name of centre

Town

Forename(s)

Surname

Date of birth

Day	Month	Year	Scottish candidate number	Number of seat

1 All questions should be attempted.

2 Necessary data will be found in the Data Booklet provided for Chemistry at Standard Grade and Intermediate 2.

3 The questions may be answered in any order but all answers are to be written in this answer book, and must be written clearly and legibly in ink.

4 Rough work, if any should be necessary, as well as the fair copy, is to be written in this book.

 Rough work should be scored through when the fair copy has been written.

5 Additional space for answers and rough work will be found at the end of the book.

6 The size of the space provided for an answer should not be taken as an indication of how much to write. It is not necessary to use all the space.

7 Before leaving the examination room you must give this book to the invigilator. If you do not, you may lose all the marks for this paper.

SCOTTISH
QUALIFICATIONS
AUTHORITY

PART 1

In Questions 1 to 9 of this part of the paper, an answer is given by circling the appropriate letter (or letters) in the answer grid provided.

In some questions, two letters are required for full marks.

If more than the correct number of answers is given, marks will be deducted.

A total of 20 marks is available in this part of the paper.

SAMPLE QUESTION

A CH$_4$	B H$_2$	C CO$_2$
D CO	E C$_2$H$_5$OH	F C

(a) Identify the hydrocarbon.

Ⓐ	B	C
D	E	F

The correct answer to part (a) is A. This should be circled.

(b) Identify the **two** elements.

A	Ⓑ	C
D	E	Ⓕ

As indicated in this question, there are **two** correct answers to part (b). These are B and F. Both answers are circled.

If, after you have recorded your answer, you decide that you have made an error and wish to make a change, you should cancel the original answer and circle the answer you now consider to be correct. Thus, in part (a), if you want to change an answer A to an answer D, your answer sheet would look like this:

A̶	B	C
Ⓓ	E	F

If you want to change back to an answer which has already been scored out, you should enter a tick (✓) in the box of the answer of your choice, thus:

✓A̶	B	C
D̶	E	F

DO NOT WRITE IN THIS MARGIN

Marks KU PS

1. The grid shows the formulae of six oxides.

A H_2O	B NO_2	C K_2O
D CaO	E CO	F SO_2

(*a*) Identify the oxide produced by the sparking of air in car engines.

A	B	C
D	E	F

1

(*b*) Identify the **two** oxides produced by burning hydrocarbons.

A	B	C
D	E	F

½

1

(2)

[Turn over

Marks KU PS

2. The names of some hydrocarbons are shown in the grid.

A cyclobutane	B cyclopentane	C butane
D propane	E ethane	F butene

(a) Identify the hydrocarbon which is a liquid at 25 °C.

You may wish to use the data booklet to help you.

A	B	C
D	E	F

✗

1

(b) Identify the **two** isomers.

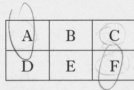

A	B	C
D	E	F

1

(c) Identify the hydrocarbon that reacts quickly with bromine solution.

A	B	C
D	E	F

1

(3)

Marks | KU | PS

3. The grid shows the names of some soluble compounds.

A	B	C
magnesium bromide	sodium bromide	lithium hydroxide
D	**E**	**F**
sodium iodide	potassium sulphate	lithium chloride

(*a*) Identify the base.

A	B	C
D	E	F

1

(*b*) Identify the **two** compounds whose solutions would form a precipitate when mixed.

You may wish to use the data booklet to help you.

A	B	C
D	E	F

1

(*c*) Identify the compound with a formula of the type **XY₂**, where **X** is a metal.

A	B	C
D	E	F

1

(3)

[Turn over

Marks | KU | PS

4. A pupil carried out the following experiments.

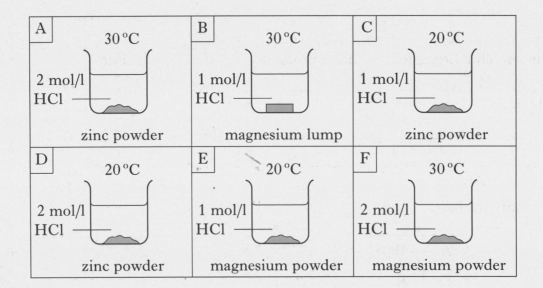

(*a*) Identify the **two** experiments which can be used to investigate the effect of concentration on the rate of the reaction.

A	B	C
D	E	F

1

(*b*) Identify the experiment with the fastest reaction rate.

A	B	C
D	E	F

1
(2)

Marks | KU | PS

5. The table contains information about some substances.

Substance	Melting point/°C	Boiling point/°C	Conducts as	
			a solid	a liquid
A	455	1567	no	yes
B	80	218	no	no
C	1492	2897	yes	yes
D	1407	2357	no	no
E	645	1287	no	yes
F	98	890	yes	yes

(*a*) Identify the **two** ionic compounds.

A
B
C
D
E
F

1

(*b*) Identify the substance which exists as a covalent network.

A
B
C
D
E
F

1

(2)

[Turn over

6. The grid shows information about some particles.

A $^{34}_{16}S^{2-}$	B $^{24}_{12}Mg^{2+}$	C $^{39}_{19}K$
D $^{40}_{19}K$	E $^{40}_{20}Ca$	F $^{35}_{17}Cl^-$

(a) Identify the **two** particles which are isotopes.

A	B	C
D	E	F

1

(b) Identify the **two** particles with the same electron arrangement as argon.

A	B	C
D	E	F

1

(2)

DO NOT WRITE IN THIS MARGIN

Marks KU PS

7. The grid contains information about the particles found in atoms.

A	B	C
charge = zero	relative mass almost zero	charge = 1−
D	**E**	**F**
found inside the nucleus	charge = 1+	relative mass = 1

Identify the **two** terms which can be applied to electrons.

A	B	C
D	E	F

(2)

[Turn over

DO NOT WRITE IN THIS MARGIN

Marks KU PS

8. Glucose, sucrose and starch are carbohydrates.

Identify the **two** correct statements.

A	Glucose molecules join together with the loss of water.
B	Starch is a polymer made from sucrose molecules.
C	Sucrose turns warm Benedict's solution orange.
D	Glucose is an isomer of sucrose.
E	Starch dissolves easily in water.
F	Sucrose can be hydrolysed.

A
B
C
D
E
F

(2)

Marks | KU | PS

9. The diagram shows how an object can be coated in nickel.

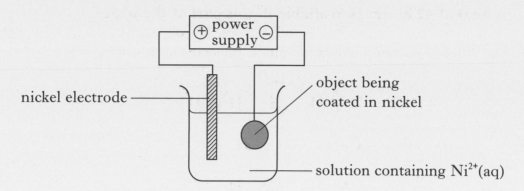

The following reactions take place at the electrodes.

Negative electrode: $Ni^{2+}(aq) + 2e^- \rightarrow Ni(s)$

Positive electrode: $Ni(s) \rightarrow Ni^{2+}(aq) + 2e^-$

Identify the **two** correct statements.

A	Nickel ions move towards the nickel electrode.
B	The mass of the nickel electrode decreases.
C	The process is an example of galvanising.
D	Oxidation occurs at the nickel electrode.
E	Electrons flow through the solution.

A
B
C
D
E

(2)

[Turn over

Marks | KU | PS

PART 2

A total of 40 marks is available in this part of the paper.

10.   The structure of part of a polyacrylonitrile molecule is shown below.

$$\sim\sim \overset{\displaystyle \overset{H}{|}}{\underset{\displaystyle \underset{H}{|}}{C}} - \overset{\displaystyle \overset{H}{|}}{\underset{\displaystyle \underset{CN}{|}}{C}} - \overset{\displaystyle \overset{H}{|}}{\underset{\displaystyle \underset{H}{|}}{C}} - \overset{\displaystyle \overset{H}{|}}{\underset{\displaystyle \underset{CN}{|}}{C}} - \overset{\displaystyle \overset{H}{|}}{\underset{\displaystyle \underset{H}{|}}{C}} - \overset{\displaystyle \overset{H}{|}}{\underset{\displaystyle \underset{CN}{|}}{C}} \sim\sim$$

(*a*)   Draw the structural formula for the monomer used to make polyacrylonitrile.

(*b*)   Name a toxic gas produced when polyacrylonitrile burns.

_____

1

1

(2)

Marks | KU | PS

11. Some Euro coins are made from a hard-wearing alloy called Nordic Gold.

(a) What is an alloy?

_____

_____   1

(b) The composition of Nordic Gold is shown in the table.

Metal	copper	aluminium	zinc	tin
% by mass	89	5	5	1

One of the coins has a mass of 5·74 g.

(i) Calculate the mass, in grams, of aluminium in the coin.
Show your working clearly.

_____ g   1

(ii) Calculate the number of moles of aluminium in the coin.
Show your working clearly.

$m = v \times c$

$=$

_____ mol   1

(3)

[Turn over

DO NOT
WRITE IN
THIS
MARGIN

Marks | KU | PS

12. A group of pupils investigated the speed of reaction between marble chips (calcium carbonate) and hydrochloric acid, concentration 1 mol/l.

They used excess hydrochloric acid to make sure all the calcium carbonate had been used up.

The pupils used a balance to measure the mass lost during the reaction.

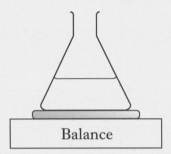

Balance

The results are shown in the table.

Time/minutes	0	0·5	1·0	2·0	3·0	4·0	5·0
Mass lost/g	0	0·30	0·50	0·70	0·76	0·79	0·80

(a) Why is mass lost during the reaction?

_____

_____

1

Marks | KU | PS

12. (continued)

(b) Draw a line graph of the results.

Use appropriate scales to fill most of the graph paper.

(Additional graph paper, if required, will be found on page 26.)

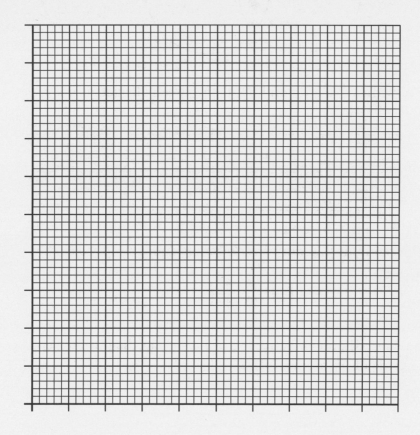

2

(c) The experiment was repeated using the same volume of hydrochloric acid, but with a concentration of 2 mol/l.

What mass loss would have been recorded?

_____ g

1

(d) Name the salt produced when marble chips react with hydrochloric acid.

_____

1

(5)

[**Turn over**

Marks | KU | PS

13. In a hydrogen molecule the atoms share two electrons in a covalent bond.

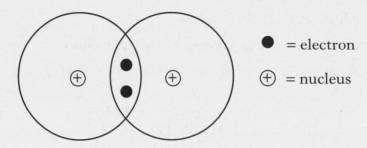

= electron

= nucleus

(a) Explain how the covalent bond holds the two hydrogen atoms together.

_____

_____

_____   1

(b) The hydrogen molecule can be represented more simply as

H $\bullet$ H

(i) Showing **all** outer electrons, draw a similar diagram to represent a molecule of ammonia, NH_3.

1

(ii) Draw another diagram to show the **shape** of an ammonia molecule.

1

(3)

Marks | KU | PS

14. Methanol can take part in many chemical reactions.

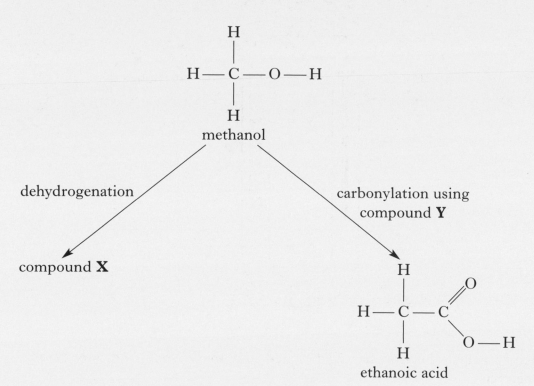

methanol

dehydrogenation

carbonylation using
compound **Y**

compound **X**

ethanoic acid

(*a*) (i) Compound **X** has the molecular formula CH_2O.
Draw the full structural formula for compound **X**.

1

(ii) Methanol is changed to compound **X** by dehydrogenation.
Suggest what is meant by **dehydrogenation**.

1

(*b*) In carbonylation, methanol reacts with compound **Y** forming only ethanoic acid.

Suggest a name for compound **Y**.

1

(3)

15. Gemma and Laura set up the simple cell shown below.

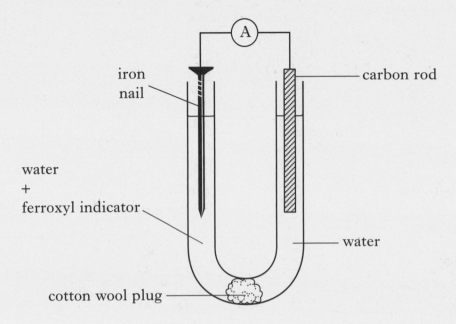

After two days the iron nail had rusted and the ferroxyl indicator had turned blue.

(a) **On the diagram**, clearly mark the path and direction of electron flow.

1

(b) The reaction taking place at the carbon rod produces hydroxide ions. How could Gemma and Laura have shown that hydroxide ions were present?

_____

_____

1

(2)

Marks | KU | PS

16. Electrolysis is a common industrial process. Some uses of electrolysis are shown in the diagram.

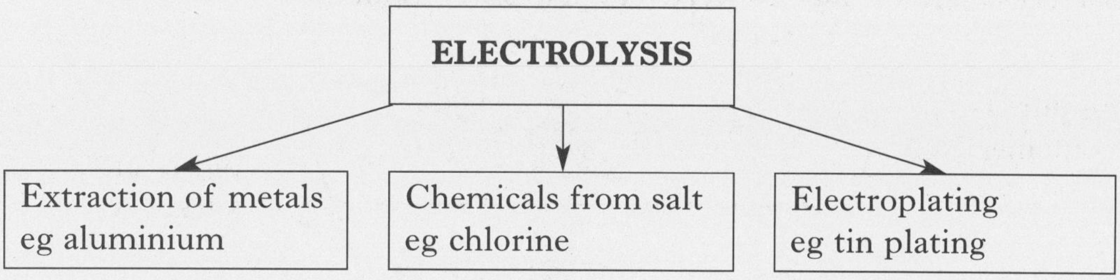

```
                    ┌─────────────────┐
                    │   ELECTROLYSIS  │
                    └─────────────────┘
         ┌───────────────┬─────────────┬──────────────┐
```

| Extraction of metals eg aluminium | Chemicals from salt eg chlorine | Electroplating eg tin plating |

(a) State what is meant by electrolysis.

_____

_____

_____

1

(b) Aluminium is extracted by electrolysis of its molten oxide since aluminium oxide does not react when heated with carbon.

Why does aluminium oxide **not** react with hot carbon?

_____

_____

_____

1

(c) Chlorine is produced by the electrolysis of sodium chloride solution.

Write the ion-electron equation for the formation of chlorine.

You may wish to use the data booklet to help you.

_____

1

(d) Tin plated iron rusts very rapidly if the plating is scratched.

Explain why the iron rusts so rapidly.

_____

_____

_____

2

(5)

[Turn over

Marks | KU | PS

17. Silver jewellery slowly tarnishes in air. This is due to the formation of silver(I) sulphide, Ag_2S.

The silver(I) sulphide can be converted back to silver as follows.

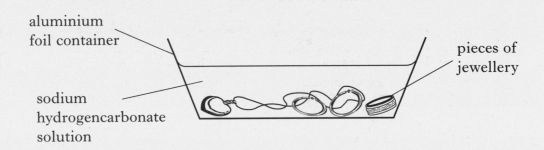

aluminium
foil container

pieces of
jewellery

sodium
hydrogencarbonate
solution

(a) Write the ionic formula for sodium hydrogencarbonate.
You may wish to use the data booklet to help you.

_____

1

(b) The equation for the reaction which takes place in the aluminium container is:

$$Ag_2S \ + \ Al \ \rightarrow \ Ag \ + \ Al_2S_3$$

(i) Balance this equation.

1

(ii) Name the type of chemical reaction which takes place.

_____

1

(c) Calculate the percentage by mass of aluminium in Al_2S_3.
Show your working clearly.

2

(5)

Marks | KU | PS

18. Fritz was investigating the properties of ammonia.

Before   **After**

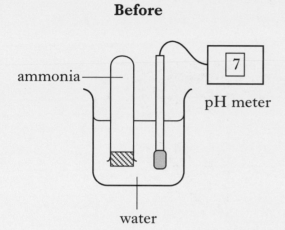 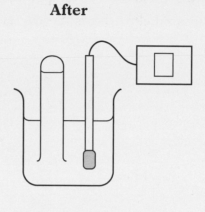

(a) Why did the water rise up the test tube when the stopper was removed?

_____

_____   1

(b) When the stopper was removed the reading on the pH meter changed. Suggest what the new reading would have been.

_____   1

       (2)

[Turn over

Marks | KU | PS

19. Water can exist in three different states: solid, liquid and gas.

The state depends on the temperature and pressure.

The diagram below shows these relationships.

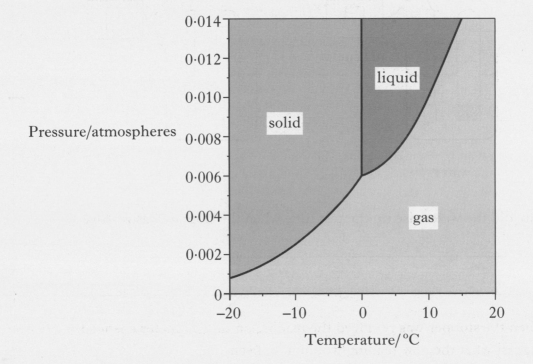

(*a*) In which state would water exist at 15 °C and 0·007 atmospheres?

_____

1

(*b*) Solid water at 0·004 atmospheres is allowed to warm up. The pressure is kept constant.

At what temperature would the solid water change into a gas?

_____ °C

1

(2)

20. Dienes are a homologous series of hydrocarbons which contain two double bonds per molecule.

buta-1,3-diene

penta-1,3-diene

hexa-1,3-diene

(a) What is meant by the term "homologous series"?

_____

_____

_____   1

(b) Suggest a general formula for the dienes.

_____   1

(c) Write the **molecular formula** for the product of the complete reaction of penta-1,3-diene with bromine.

_____   1

(d) Draw a full structural formula for an isomer of buta-1,3-diene which contains only **one** double bond per molecule.

1

(4)

Marks KU PS

21. A pupil carried out a titration using the chemicals and apparatus shown below.

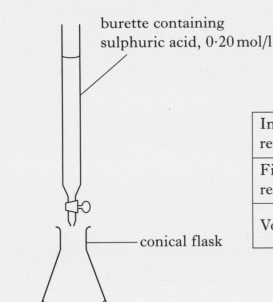

burette containing
sulphuric acid, 0·20 mol/l

conical flask

20 cm³ potassium hydroxide solution + indicator

	Rough titre	1st titre	2nd titre
Initial burette reading/cm³	0·5	21·7	0·3
Final burette reading/cm³	21·7	42·4	20·8
Volume used/cm³	21·2	20·7	20·5

(a) How would the pupil know when to stop adding acid from the burette?

_____

_____ 1

(b) (i) What average volume should be used to calculate the number of moles of sulphuric acid needed to neutralise the potassium hydroxide solution?

_____ cm³ 1

DO NOT WRITE IN THIS MARGIN

Marks | KU | PS

21. (*b*) (continued)

(ii) Calculate the number of moles of sulphuric acid in this average volume.

Show your working clearly.

__________ mol 1

(iii) The equation for the titration reaction is

$$H_2SO_4 \quad + \quad 2KOH \quad \rightarrow \quad K_2SO_4 \quad + \quad 2H_2O$$

Calculate the number of moles of potassium hydroxide in 20 cm^3 of the potassium hydroxide solution.

Show your working clearly.

1mol 2mol

0·0412 0·0824

__________ mol 1

(4)

[*END OF QUESTION PAPER*]

ADDITIONAL SPACE FOR ANSWERS

ADDITIONAL GRAPH PAPER FOR QUESTION 12(*b*)

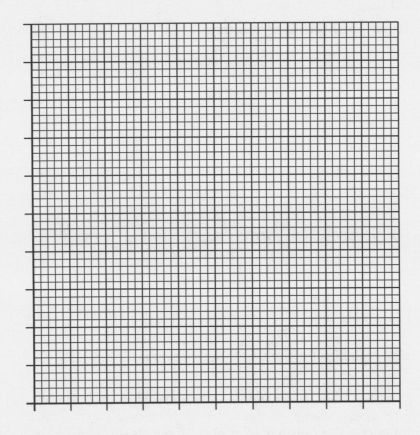

BLANK PAGE

C

FOR OFFICIAL USE

KU   PS

Total
Marks

0500/402

NATIONAL
QUALIFICATIONS
2005

MONDAY, 9 MAY
10.50 AM – 12.20 PM

CHEMISTRY
STANDARD GRADE
Credit Level

Fill in these boxes and read what is printed below.

Full name of centre

Town

Forename(s)

Surname

Date of birth
Day   Month   Year   Scottish candidate number   Number of seat

1   All questions should be attempted.

2   Necessary data will be found in the Data Booklet provided for Chemistry at Standard Grade and Intermediate 2.

3   The questions may be answered in any order but all answers are to be written in this answer book, and must be written clearly and legibly in ink.

4   Rough work, if any should be necessary, as well as the fair copy, is to be written in this book.

Rough work should be scored through when the fair copy has been written.

5   Additional space for answers and rough work will be found at the end of the book.

6   The size of the space provided for an answer should not be taken as an indication of how much to write. It is not necessary to use all the space.

7   Before leaving the examination room you must give this book to the invigilator. If you do not, you may lose all the marks for this paper.

SCOTTISH
QUALIFICATIONS
AUTHORITY

SAB 0500/402 6/28670   ©

PART 1

In Questions 1 to 9 of this part of the paper, an answer is given by circling the appropriate letter (or letters) in the answer grid provided.

In some questions, two letters are required for full marks.

If more than the correct number of answers is given, marks will be deducted.

A total of 20 marks is available in this part of the paper.

SAMPLE QUESTION

A		B		C	
	CH_4		H_2		CO_2
D		E		F	
	CO		C_2H_5OH		C

(a) Identify the hydrocarbon.

Ⓐ	B	C
D	E	F

The one correct answer to part (a) is A. This should be circled.

(b) Identify the **two** elements.

A	Ⓑ	C
D	E	Ⓕ

As indicated in this question, there are **two** correct answers to part (b). These are B and F. Both answers are circled.

If, after you have recorded your answer, you decide that you have made an error and wish to make a change, you should cancel the original answer and circle the answer you now consider to be correct. Thus, in part (a), if you want to change an answer A to an answer D, your answer sheet would look like this:

Ⓐ̸	B	C
Ⓓ	E	F

If you want to change back to an answer which has already been scored out, you should enter a tick (✓) in the box of the answer of your choice, thus:

✓Ⓐ̸	B	C
Ⓓ̸	E	F

Marks | KU | PS

1. The grid shows the names of some elements.

A	B	C
argon	potassium	magnesium
D	E	F
chlorine	phosphorus	sulphur

(a) Identify the element which produces a lilac flame colour.

You may wish to use the data booklet to help you.

A	B	C
D	E	F

1

(b) Identify the element with atoms which have the same electron arrangement as a Ca^{2+} ion.

A	B	C
D	E	F

1

(c) Identify the **two** elements which would form a covalent compound with a formula of the type X_3Y_2.

A	B	C
D	E	F

1

(3)

[Turn over

× ×

S

V 2¹ 3

S 2 3

D 3 2

F 3 2

Marks

2. Distillation of crude oil produces several fractions.

Fraction	Number of carbon atoms per molecule
A	1–4
B	4–10
C	10–16
D	16–20
E	20+

Crude oil ⟶

(a) Identify the fraction which is used as a fuel in camping gas stoves.

A
B
C
D
E

(b) Identify the fraction with the highest viscosity.

A
B
C
D
E

1

1

(2)

DO NOT WRITE IN THIS MARGIN

Marks KU PS

3. The names of some carbohydrates are shown.

A	glucose
B	fructose
C	maltose
D	sucrose
E	starch

(*a*) Identify the carbohydrate which does not dissolve well in water.

A
B
C
D
E

1

(*b*) Identify the **two** carbohydrates which are disaccharides.

A
B
C
D
E

1

(2)

[Turn over

4. The table contains information about some substances.

Substance	Melting point/°C	Boiling point/°C	Conducts as a solid	Conducts as a liquid
A	−7	59	no	no
B	98	883	yes	yes
C	−39	357	yes	yes
D	547	1265	no	yes
E	−78	−33	no	no
F	1700	2230	no	no

(a) Identify the substance which is a gas at 0 °C.

solid liquid gas

A
B
C
D
(E)
F

1

(b) Identify the **two** substances which exist as molecules.

(A)
B
C
D
(E)
F

1

(2)

Marks | KU | PS

5. The grid shows the formulae of some hydrocarbons.

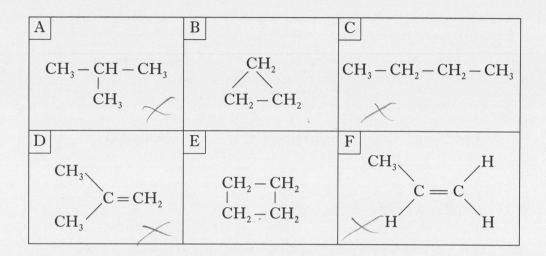

(a) Identify the hydrocarbon which can be used to make poly(propene).

A	B	C
D	E	F

1

(b) Identify the **two** hydrocarbons with the general formula C_nH_{2n} which do **not** react quickly with bromine solution.

A	B	C
D	E	F

1

(c) Identify the **two** isomers of

$$CH_2 = CH - CH_2 - CH_3$$

A	B	C
D	E	F

2

(4)

[Turn over

6. There are many different types of chemical reaction.

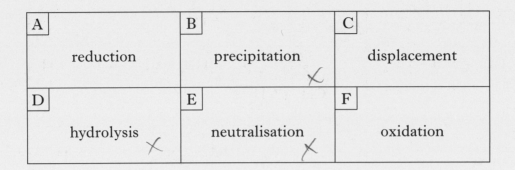

A reduction	B precipitation ✗	C displacement
D hydrolysis ✗	E neutralisation ✗	F oxidation

Identify the following type of reaction.

$$SO_3^{2-}(aq) \ + \ H_2O(\ell) \ \longrightarrow \ SO_4^{2-}(aq) \ + \ 2H^+(aq) \ + \ 2e^-$$

Ⓐ	B	C
D	E	F

(1)

7. The grid shows some statements which could be applied to a solution.

A	It does not react with magnesium.
B	It has a pH less than 7.
C	It does not conduct electricity.
D	It produces chlorine gas when electrolysed.
E	It contains more H^+ ions than pure water.

Identify the **two** statements which are true for **both** dilute hydrochloric acid and dilute sulphuric acid.

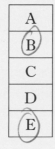

(2)

[Turn over

Marks | KU | PS

8. David was studying the reactions of some metals and their compounds.

He carried out experiments involving magnesium, copper, zinc, nickel, silver and an unknown metal **X**.

Listed below are some of the observations he recorded.

A	**X** was more readily oxidised than copper.
B	**X** oxide was more stable to heat than silver oxide.
C	Magnesium displaced **X** from a solution of **X** nitrate.
D	**X** reacted more vigorously than nickel with dilute acid.
E	Compounds of **X** were more readily reduced than compounds of zinc.

From his observations, David produced the following order of reactivity.

magnesium, zinc, nickel, copper, **X**, silver

decreasing activity

Identify the **two** observations which can be used to show that **X** has been wrongly placed.

A
B
C
D
E

(2)

Marks | KU | PS

9. Equations are used to represent chemical reactions.

A	$H^+(aq) + OH^-(aq) \rightarrow H_2O(\ell)$
B	$Fe^{3+}(aq) + e^- \rightarrow Fe^{2+}(aq)$
C	$Fe(s) \rightarrow Fe^{2+}(aq) + 2e^-$
D	$Fe^{2+}(aq) + 2e^- \rightarrow Fe(s)$
E	$H_2(g) \rightarrow 2H^+(aq) + 2e^-$
F	$2H_2O(\ell) + O_2(g) + 4e^- \rightarrow 4OH^-(aq)$

Identify the **two** equations which are involved in the corrosion of iron.

A
B
C
D
E
F

(2)

[Turn over

Page eleven

[Turn over for Question 10 on *Page thirteen*

DO NOT
WRITE IN
THIS
MARGIN

Marks | KU | PS

PART 2

A total of 40 marks is available in this part of the paper.

10. On some boats the steel propellers have zinc blocks attached to help prevent rusting. The zinc is oxidised, protecting the steel.

zinc block   steel propeller

(a) (i) Write the ion-electron equation for the oxidation of zinc. You may wish to use the data booklet to help you.

$$Zn(s) \longrightarrow Zn^{2+}(aq) + 2e^-$$

①

(ii) What name is given to the **type** of protection provided by the zinc?

sacrificial

①

(b) If cobalt is used instead of zinc the steel propeller rusts quickly. What does this suggest about the reactivity of cobalt compared to iron?

The reactivity is less — cobalt lower in series

①
(3)

[Turn over

Marks | KU | PS

11. Polystyrene is an addition polymer. It is made from the monomer styrene.

$$\begin{array}{ccc} H & & H \\ | & & | \\ C & = & C \\ | & & | \\ H & & C_6H_5 \end{array}$$

styrene

(a) Draw a section of the polystyrene structure, showing three monomer units joined together.

$$\begin{array}{ccccccccccc} H & & H & & H & & H & & H & & H \\ | & & | & & | & & | & & | & & | \\ C & - & C & - & C & - & C & - & C & - & C \\ | & & | & & | & & | & & | & & | \\ H & & C_6H_5 & & H & & C_6H_3 & & H & & C_6H_3 \end{array}$$

1

(b) Calculate the percentage by mass of carbon in a molecule of styrene.

$$\% \ mass = \frac{mass \ of \ element}{gram \ formula \ mass}$$

$$= 12$$

Answer _____ %

2

(3)

DO NOT
WRITE IN
THIS
MARGIN

Marks | KU | PS

12. Methane (CH_4), ethane (C_2H_6) and propane (C_3H_8) are the first three members of the alkanes.

(*a*) State the general formula for the alkanes.

<u>$C_n H_{2n+2}$</u>

1

(*b*) The ninth member of the alkanes is nonane (C_9H_{20}).

 (i) Predict the boiling point of nonane.
   You may wish to use page 6 of the data booklet to help you.

   <u>~~124~~ 147</u> °C C_8H_{18}

   ✗

 (ii) Nonane can be cracked to produce smaller, more useful hydrocarbons. A catalyst is used to speed up this reaction.
   Suggest another reason for using a catalyst.

   <u>Allows the plant lower temps</u>

   1

(*c*) Alkanes can be made by the reaction of sodium with iodoalkanes.
For example, butane can be made from iodoethane.

iodoethane   iodoethane   butane

Butane can also be made using two **different** iodoalkanes.
Name the **two** iodoalkanes.

<u>iopropane, iomethane</u>

1

(4)

[Turn over

Official SQA Past Papers: Credit Chemistry 2005

DO NOT
WRITE IN
THIS
MARGIN

KU | PS

13. A mass spectrometer is an instrument that can be used to measure the percentage of isotopes in a sample of an element.

When a sample of chlorine is passed through a mass spectrometer the following graph is obtained.

Each spike on the graph shows the presence of an isotope.

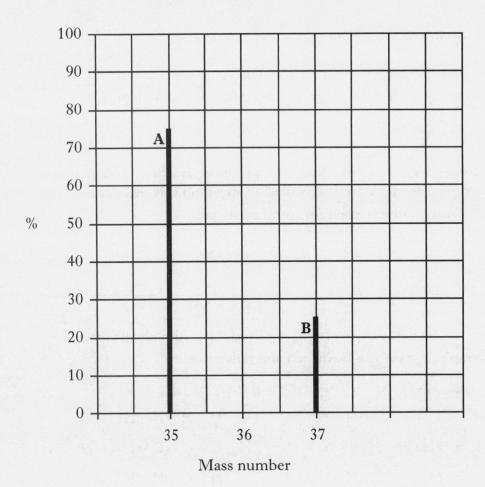

Mass number

The **relative atomic mass** of an element can be calculated using the formula:

$$\frac{(\text{mass of isotope } \mathbf{A} \times \%) + (\text{mass of isotope } \mathbf{B} \times \%)}{100}$$

The **relative atomic mass** of chlorine $= \dfrac{(35 \times 75) + (37 \times 25)}{100}$

$= 35 \cdot 5$

DO NOT WRITE IN THIS MARGIN

13. (continued)

(*a*) The following graph was obtained for a sample of lithium.

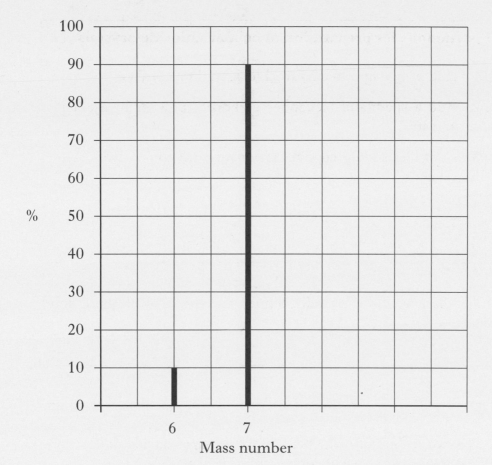

%

Mass number

(i) How many isotopes are present in the sample of lithium?

_____

1

(ii) Using the information in the graph calculate the relative atomic mass of lithium.

Show your working clearly.

1

(*b*) Complete the table to show the number of each type of particle in the ion, $_3^7Li^+$.

Particle	Number
proton	3
neutron	4
electron	2

2

(4)

Marks | KU | PS

14. Clare carried out an experiment to make copper chloride crystals.

Instructions for preparation of copper chloride crystals

Step 1 Add $25\,cm^3$ of dilute hydrochloric acid to a beaker.

Step 2 Add a spatulaful of copper carbonate powder to the acid and stir.

Step 3 Continue adding copper carbonate until some of the solid remains.

Step 4

Step 5

(*a*) Why did Clare continue to add copper carbonate until some solid remained?

To make sure it had all reacted

1

(*b*) Name the **two** techniques which Clare would have carried out in steps **4** and **5** to prepare a sample of copper chloride crystals.

Step 4   Filter solution through filter paper into beaker

Step 5   Collect filtrate + allow to crystallise

2

(3)

Marks

15. Two atoms of nitrogen share electrons to form a nitrogen molecule.

(a) Draw a diagram to show how the outer electrons are arranged in a molecule of nitrogen, N_2.

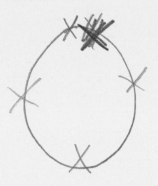

1

(b) Oxides of nitrogen dissolve in water to produce nitric acid.

   (i) Name the industrial process used to manufacture nitric acid.

       Haber

1

   (ii) A platinum catalyst is used in the industrial manufacture of nitric acid.

       Why is it **not** necessary to continue heating the platinum once the reaction has started?

1
(3)

[Turn over

Marks | KU | PS

16. Dilute hydrochloric acid reacts with sodium thiosulphate solution ($Na_2S_2O_3$) to produce a precipitate of sulphur.

$$HCl \ + \ Na_2S_2O_3 \ \rightarrow \ NaCl \ + \ S \ + \ SO_2 \ + \ H_2O$$

(*a*) Balance this equation.

1

(*b*) A pupil investigated the effect of temperature on the speed of the reaction. She measured the time taken for enough sulphur to form to make the cross disappear.

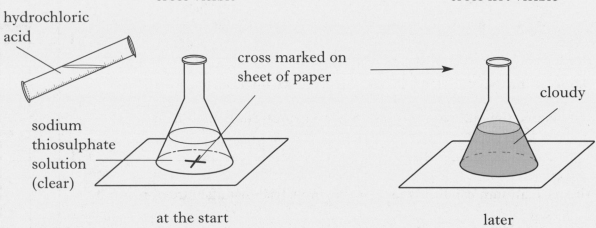

cross visible   cross not visible

hydrochloric acid

cross marked on sheet of paper

cloudy

sodium thiosulphate solution (clear)

at the start   later

Her results are shown in the table.

Temperature/°C	Time/s
25	89
30	64
35	44
40	33
45	27
50	21

Marks | KU | PS

16. *(b)* **(continued)**

(i) Draw a line graph of the results.

Use appropriate scales to fill most of the graph paper.

(Additional graph paper, if required, will be found on page 27.)

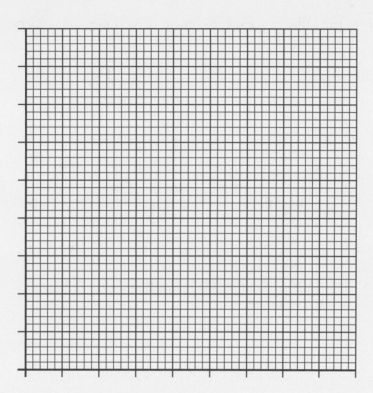

2

(ii) Use your graph to estimate the time taken, in seconds, for the cross to disappear at 60 °C.

_____

1

(iii) Describe the relationship between the temperature and the **speed** of the reaction.

_____

_____

1

(c) State **one** factor that must be kept constant throughout this investigation.

_____

1

(6)

[Turn over

Marks | KU | PS

17. Glucose and starch are both carbohydrates.

(a) Write the molecular formula for glucose.

_____ **1**

(b) A pupil set up the following experiment to turn starch into glucose.

Test tube **A**   Test tube **B**

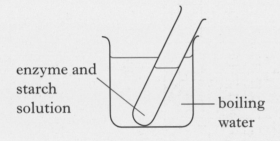
enzyme and
starch
solution — boiling
water

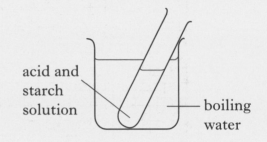
acid and
starch
solution — boiling
water

(i) Name the type of chemical reaction which takes place when starch is broken down to glucose.

_____ **1**

(ii) Suggest why glucose would **not** be formed in test tube **A**.

_____

_____

_____ **1**

(3)

Marks | KU | PS

18. Copper can be extracted from its oxide by heating copper(II) oxide with hydrogen gas. Water is also formed during the reaction.

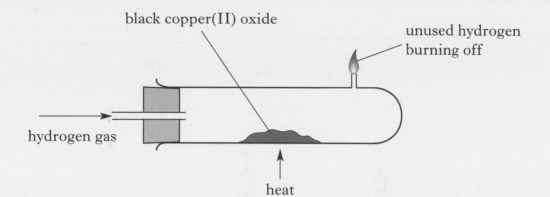

black copper(II) oxide

unused hydrogen burning off

hydrogen gas

heat

(a) Write an equation, using symbols and formulae, for the reaction between copper(II) oxide and hydrogen gas.
There is no need to balance the equation.

_____ 1

(b) Suggest the colour change which would be seen in the copper(II) oxide during the reaction.

_____ 1

(c) Suggest why calcium cannot be extracted from its oxide by heating with hydrogen gas.

_____

_____

_____ 1

(3)

[Turn over

Marks KU | PS

19. The cell below can be used in a carbon monoxide detector.

Carbon monoxide enters the cell along with oxygen from the air at electrode **A**.

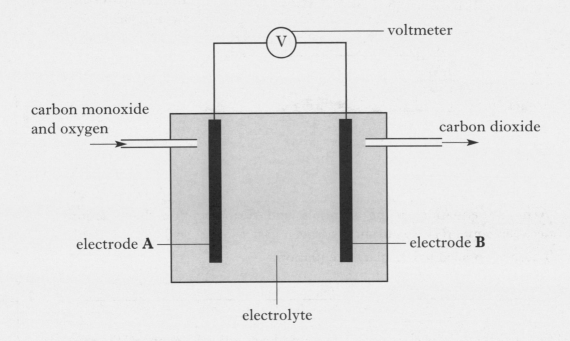

The reactions occurring at each electrode are:

Electrode A

$$CO(g) + H_2O(\ell) \longrightarrow CO_2(g) + 2H^+(aq) + 2e^-$$

Electrode B

$$2H^+(aq) + \frac{1}{2}O_2(g) + 2e^- \longrightarrow H_2O(\ell)$$

(a) **On the diagram**, clearly mark the path and direction of electron flow.   1

(b) What is the purpose of the electrolyte in the above cell?

_____

_____   1

DO NOT WRITE IN THIS MARGIN

Marks | KU | PS

19. (continued)

(*c*) Sugar solution cannot be used as an electrolyte.

What does this indicate about the bonding in sugar?

covalent

1

(*d*) Platinum is used for the electrodes in this cell.

(i) To which family of metals does platinum belong?

1

(ii) Platinum is also used as a catalyst in a catalytic converter in car exhausts.

What does a catalytic converter do?

1

(5)

[Turn over

Marks | KU | PS

20. Aluminium powder reacts with dilute sulphuric acid.

$$2Al(s) \quad + \quad 3H_2SO_4(aq) \quad \longrightarrow \quad Al_2(SO_4)_3(aq) \quad + \quad 3H_2(g)$$

(a) Circle the formula for the salt in the above equation.

1

(b) Calculate the mass of hydrogen produced when 1·35 g of aluminium reacts with dilute sulphuric acid.

2 mole
54g
·1 g
1·35g

3 mole
6g
$\frac{6}{54} \times 1·35$

Answer ___0·15___ g   2

(3)

[END OF QUESTION PAPER]

ADDITIONAL SPACE FOR ANSWERS

ADDITIONAL GRAPH PAPER FOR QUESTION 16(b)(i)

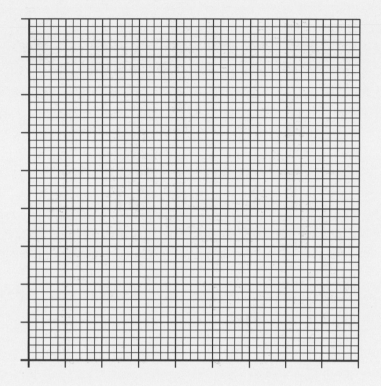

ADDITIONAL SPACE FOR ANSWERS

[BLANK PAGE]

C

FOR OFFICIAL USE

	KU	PS
Total Marks		

0500/402

NATIONAL
QUALIFICATIONS
2006

MONDAY, 8 MAY
10.50 AM – 12.20 PM

CHEMISTRY
STANDARD GRADE
Credit Level

Fill in these boxes and read what is printed below.

Full name of centre

Town

Forename(s)

Surname

Date of birth

Day Month Year   Scottish candidate number   Number of seat

1 All questions should be attempted.

2 Necessary data will be found in the Data Booklet provided for Chemistry at Standard Grade and Intermediate 2.

3 The questions may be answered in any order but all answers are to be written in this answer book, and must be written clearly and legibly in ink.

4 Rough work, if any should be necessary, as well as the fair copy, is to be written in this book.

Rough work should be scored through when the fair copy has been written.

5 Additional space for answers and rough work will be found at the end of the book.

6 The size of the space provided for an answer should not be taken as an indication of how much to write.   It is not necessary to use all the space.

7 Before leaving the examination room you must give this book to the invigilator. If you do not, you may lose all the marks for this paper.

SCOTTISH
QUALIFICATIONS
AUTHORITY

PART 1

In **Questions 1 to 7** of this part of the paper, an answer is given by circling the appropriate letter (or letters) in the answer grid provided.

In some questions, two letters are required for full marks.

If more than the correct number of answers is given, marks will be deducted.

A total of 20 marks is available in this part of the paper.

SAMPLE QUESTION

A CH_4	B H_2	C CO_2
D CO	E C_2H_5OH	F C

(a) Identify the hydrocarbon.

Ⓐ	B	C
D	E	F

The one correct answer to part (a) is A. This should be circled.

(b) Identify the **two** elements.

A	Ⓑ	C
D	E	Ⓕ

As indicated in this question, there are **two** correct answers to part (b). These are B and F. Both answers are circled.

If, after you have recorded your answer, you decide that you have made an error and wish to make a change, you should cancel the original answer and circle the answer you now consider to be correct. Thus, in part (a), if you want to change an answer A to an answer D, your answer sheet would look like this:

Ⱥ	B	C
Ⓓ	E	F

If you want to change back to an answer which has already been scored out, you should enter a tick (✓) in the box of the answer of your choice, thus:

✓Ⱥ	B	C
Ⱦ	E	F

Marks | KU | PS

1. The grid shows the names of some metals.

A	B	C
potassium	platinum	iron
D	E	F
tin	copper	magnesium

(a) Identify the metal used as a catalyst in the Ostwald Process.

A	B	C
D	E	F

1

(b) Identify the metal produced in a Blast Furnace.

A	B	C⃝
D	E	F

1

(c) Identify the metal which has a density of $8 \cdot 92$ g/cm^3.

You may wish to use the data booklet to help you.

A	B	C
D	E	F

1

(3)

[Turn over

2. The structures of some hydrocarbons are shown in the grid below.

A	B	C

H   CH₃ H
 	
H−C−C—C−H
 	
H   CH₃ H
	H H
	  \ /
	   C
	  / \
	H−C——C−H
	   | |
	   H H
		H H H
		
		H−C−C−C−H
		
		H H H

D	**E**	**F**

H H H H
 	
H−C−C−C−C−H
 	
H H H H
	H H H H
	            \ /
	H−C−C−C=C
	              \
	H H H
		H H
		H−C−C−H
		
		H−C−C−H
		
		H H

(a) Identify the **two** isomers.

A	B	C
D	(E)	(F)

(b) Identify the hydrocarbon which is the first member of a homologous series.

A	(B)	C
D	E	F

1

(2)

Marks | KU | PS

3. The grid shows the formulae for some gases.

A	B	C
O_2	N_2	CO
D	E	F
SO_2	NO_2	CO_2

(*a*) Identify the poisonous gas produced during the **incomplete combustion** of hydrocarbons.

A	B	Ⓒ
D	E̶	F

1

(*b*) Identify the gas produced in air during a lightning storm.

A̶	B	C
D	Ⓔ	F

1

(*c*) Identify the gas which is a reactant in the manufacture of ammonia (Haber Process).

A	Ⓑ	C
D	E	F

1

(3)

[Turn over

DO NOT WRITE IN THIS MARGIN

Marks KU PS

4. The names of some oxides are shown in the grid.

A sodium oxide	B potassium oxide	C copper(II) oxide
D carbon dioxide	E zinc oxide	F sulphur dioxide

(*a*) Identify the **two** oxides which dissolve in water to form alkaline solutions.

A	B	C
D	E	F

1

(*b*) Identify the **two** oxides which are covalent.

A	B	C
D	E	F

1

(2)

Marks KU PS

5. The Periodic Table lists all known elements.

The grid shows the names of six common elements.

A oxygen	B calcium	C aluminium
D sodium	E magnesium	F fluorine

(a) Identify the **two** elements with similar chemical properties.

A	B	C
D	E	F

1

(b) Identify the element which can form ions with the same electron arrangement as argon.

A	B	C
D	E	F

1

(c) Identify the **two** elements which form an ionic compound with the formula of the type **XY₃**, where **X** is the metal.

A	B	C
D	E	F

1

(3)

[Turn over

Marks | KU | PS

6. There are many different types of chemical reaction.

A precipitation	B hydrolysis	C oxidation
D neutralisation	E condensation	F addition

(a) Identify the type of chemical reaction that occurs when ethene reacts with hydrogen to form ethane.

A	B	C
D	E	Ⓕ

1

(b) Identify the type of chemical reaction which occurs when a metal corrodes.

A	B	Ⓒ
D	E	F

1

(c) Identify the **two** types of chemical reaction represented by the following equation.

$$Ba^{2+}(OH^-)_2(aq) + (H^+)_2SO_4^{2-}(aq) \longrightarrow Ba^{2+}SO_4^{2-}(s) + 2H_2O(\ell)$$

A	B	C
D	E	F

2

(4)

Page eight

DO NOT WRITE IN THIS MARGIN

Marks KU PS

7. The grid shows pairs of chemicals.

A	B
$CuO + C$	$Na + H_2O$
C	D
$Cu + NaNO_3$	$C_5H_{12} + O_2$
E	F
$Mg + H_2SO_4$	$Ag + HCl$

(*a*) Which box contains a pair of chemicals that react to form water?

A	B
C	D
E	F

1

(*b*) Which **two** boxes contain pairs of chemicals that do **not** react together?

A	B
C	D
E	F

2

(3)

[Turn over

DO NOT
WRITE IN
THIS
MARGIN

Marks | KU | PS

PART 2

A total of 40 marks is available in this part of the paper.

8. Teflon is the brand name for the plastic, poly(tetrafluoroethene).

   The structure of part of a poly(tetrafluoroethene) molecule is shown below.

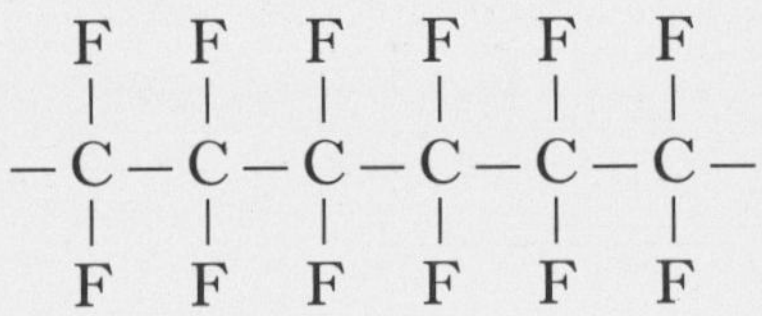

$$\begin{array}{cccccc} F & F & F & F & F & F \\ | & | & | & | & | & | \\ -C- & C- & C- & C- & C- & C- \\ | & | & | & | & | & | \\ F & F & F & F & F & F \end{array}$$

   (*a*) Draw the full structural formula for the monomer used to make poly(tetrafluoroethene).

1

   (*b*) Teflon is a plastic which melts on heating.

   What name is given to this type of plastic?

   _____

1

(2)

Marks | KU | PS

9. A sample of hydrogen was found to contain two different types of atom; $_1^2H$ and $_1^1H$.

(a) (i) What term is used to describe these different types of hydrogen atom?

_____Isotope_____ **1**

(ii) This sample of hydrogen has an average atomic mass of 1·1.

What is the mass number of the most common type of atom in this sample of hydrogen?

_____1_____ **1**

(iii) Complete the table to show the number of protons and neutrons in each type of hydrogen atom.

Type of atom	Number of protons	Number of neutrons
$_1^2H$	1	1
$_1^1H$	1	0

1

(b) In a methane molecule (CH_4), hydrogen atoms form bonds with a carbon atom.

Draw a diagram to show the **shape** of a methane molecule.

1

(4)

[Turn over

Marks KU PS

10. A pupil set up the following experiment.

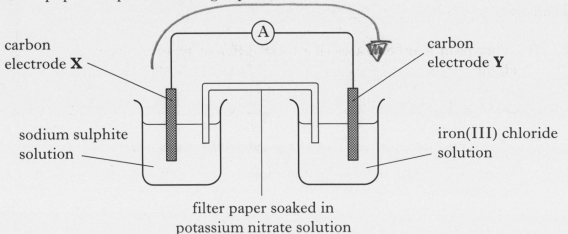

carbon electrode **X**

carbon electrode **Y**

sodium sulphite solution

iron(III) chloride solution

filter paper soaked in
potassium nitrate solution

The reaction occurring at electrode **Y** is

$$Fe^{3+}(aq) + e^- \longrightarrow Fe^{2+}(aq)$$

(a) **On the diagram**, clearly mark the path and direction of electron flow.

1

(b) Name the type of chemical reaction taking place at electrode **Y**.

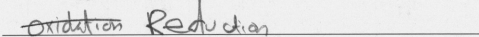

~~Oxidation~~ Reduction

1

(c) After some time, ferroxyl indicator was added to the beaker containing electrode **Y**.

What colour would the ferroxyl indicator turn?

1

(3)

Marks | KU | PS

11. Some types of steel can rust.

(a) Name the **two** substances which must be present for steel to rust.

_____ **1**

(b) Paint containing "Red Lead" (Pb_3O_4) was used to protect steel from rusting.

Calculate the percentage, by mass, of lead in "Red Lead".

$$\% = \frac{mass\,Pb}{GFM} \times 100$$

$$= \frac{621}{}$$

Pb_3O_4

_____ % **2**

(c) Stainless steel is a type of steel which does not need protection. It contains chromium which forms a layer of chromium(III) oxide on the steel.

Write the formula for chromium(III) oxide.

_____ **1**

(4)

```
S   Cr   O
V   3    2
S        X
D
V
```

[Turn over

12. Laura added the catalyst manganese dioxide to hydrogen peroxide solution and measured the volume of oxygen produced.

$$2H_2O_2(aq) \longrightarrow 2H_2O\ (\ell) + O_2(g)$$

Her results are shown in the table.

Time/s	0	10	30	40	50	60
Volume of oxygen/cm^3	0	25	35	38	40	40

(a) Draw a line graph of the results.

Use appropriate scales to fill most of the graph paper.

(Additional graph paper, if required, will be found on page 25.)

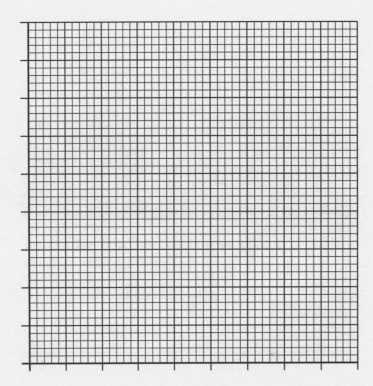

2

DO NOT WRITE IN THIS MARGIN

Marks KU PS

12. (continued)

(*b*) Using your graph, predict the volume of oxygen produced during the first 20 seconds.

_____________ cm^3 **1**

(*c*) Laura repeated the experiment at a higher temperature. She used the same volume and concentration of hydrogen peroxide solution.

Suggest a volume of oxygen produced during the first 30 seconds.

_____________ cm^3 **1**

(4)

[Turn over

13. Glucose is a carbohydrate.

(a) Name an isomer of glucose.

_____

1

(b) Glucose molecules join together to form starch in a polymerisation reaction.

Name the **type** of polymerisation reaction which takes place.

_____

1

(c) The diagram shows how glucose can be fermented to produce an alcohol. Carbon dioxide gas is also produced.

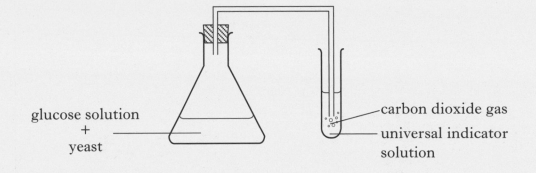

glucose solution
+
yeast

carbon dioxide gas
universal indicator
solution

(i) What is the chemical name for the alcohol produced?

_____

1

(ii) Suggest the colour of the universal indicator solution after the carbon dioxide gas has been bubbled through it.

_____

1

(4)

Page sixteen

14. Fuel cells produce electricity to power cars. The electricity is produced when hydrogen and oxygen react to form water.

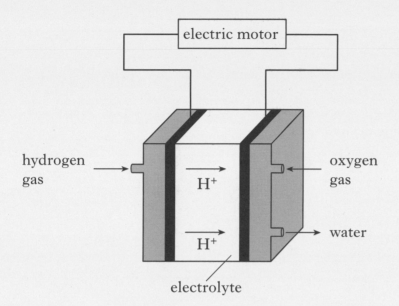

(a) Suggest a possible source of oxygen for use in the fuel cell.

_____

1

(b) Suggest an advantage in using fuel cells rather than petrol to power cars.

_____

_____

1

(c) Write the ion-electron equation for the formation of hydrogen ions.

You may wish to use the data booklet to help you.

_____

1

(3)

[Turn over

Marks | KU | PS

15. Dinitrogen monoxide can be used to increase power in racing cars.

Dinitrogen monoxide decomposes to form nitrogen and oxygen.

$$2N_2O(g) \longrightarrow 2N_2(g) + O_2(g)$$

dinitrogen monoxide

(*a*) Calculate the mass of oxygen produced, in grams, when 22 grams of dinitrogen monoxide decomposes.

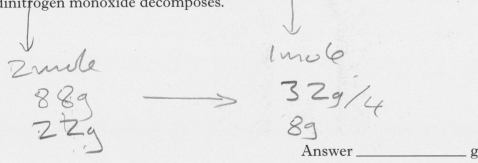

Answer _____ g   **2**

(*b*) Tom set up the following experiment to compare the time taken for the burning candles in gas jars **A** and **B** to go out.

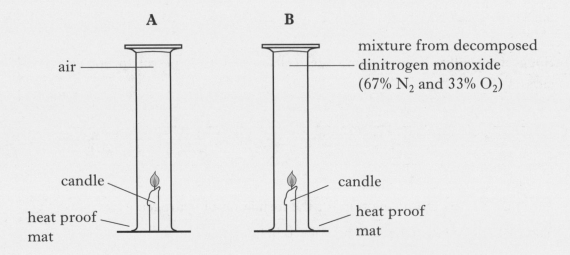

A   B

air ————   ———— mixture from decomposed
                              dinitrogen monoxide
                              (67% N_2 and 33% O_2)

candle   candle

heat proof   heat proof
mat   mat

Circle the correct word in the table to show how the burning time of the candle in gas jar **B** compared to that in gas jar **A**.

Candle	Burning time/s
A	10
B	same/longer/shorter

1

(3)

Marks | KU | PS

16. Ammonia gas is produced when barium hydroxide reacts with ammonium chloride.

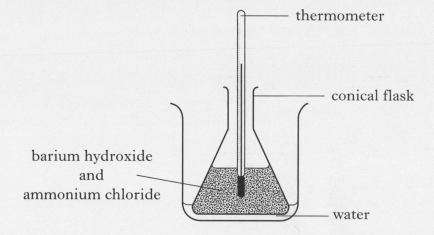

(*a*) The equation for the reaction which takes place is:

$$Ba(OH)_2 \ + \ NH_4Cl \ \rightarrow \ NH_3 \ + \ BaCl_2 \ + \ H_2O$$

Balance this equation.

1

(*b*) Describe a test which would detect ammonia at the mouth of the flask.

_____

_____

_____

1

(*c*) During the reaction the reading on the thermometer dropped from **25 °C** to **–5 °C**.

Suggest what would happen to the water in the beaker.

_____

_____

_____

1
(3)

[Turn over

Marks | KU | PS

17. A class were given three chemicals labelled **X**, **Y** and **Z**.

The chemicals were glucose solution, copper chloride solution and dilute hydrochloric acid.

The apparatus below was used to help identify each solution.

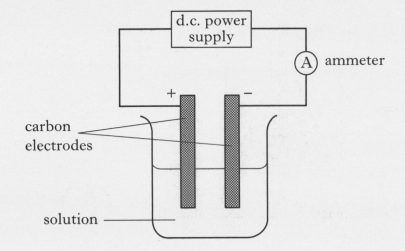

The class obtained the following results.

Solution	Ammeter reading	Observations at electrodes
X	Yes	bubbles of gas formed at both electrodes
Y	Yes	brown solid formed at negative electrode
Z	No	no reaction

(a) When electricity is passed through solutions **X** and **Y** they are broken up.

What term is used to describe this process?

_____

1

DO NOT WRITE IN THIS MARGIN

Marks KU PS

17. (continued)

(*b*) (i) Identify **X**.

______________________________ 1

(ii) What type of bonding is present in **Z**?

______________________________ 1

(iii) Describe what was **seen** at the positive electrode when electricity was passed through solution **Y**.

______________________________ 1

(4)

[Turn over

Marks | KU | PS

18. A pupil carried out a titration experiment to find the concentration of a potassium hydroxide solution.

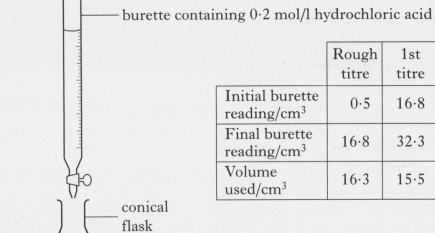

burette containing 0·2 mol/l hydrochloric acid

	Rough titre	1st titre	2nd titre
Initial burette reading/cm³	0·5	16·8	32·3
Final burette reading/cm³	16·8	32·3	48·0
Volume used/cm³	16·3	15·5	15·7

conical flask

20 cm³ potassium hydroxide solution + indicator

The equation for the reaction is:

$$KOH(aq) \quad + \quad HCl(aq) \longrightarrow KCl(aq) \quad + \quad H_2O\,(\ell)$$

(*a*) Using the results in the table, calculate the **average** volume of hydrochloric acid required to neutralise the potassium hydroxide solution.

_____ cm³   **1**

DO NOT WRITE IN THIS MARGIN

Marks | KU | PS

18. (continued)

(*b*) Use the pupil's results to calculate the concentration, in mol/l, of the potassium hydroxide solution.

Show your working clearly.

________________ mol/l **2**

(*c*) The indicator was removed from the potassium chloride solution by filtering the solution through charcoal.

How would the pupil then obtain a sample of **solid** potassium chloride from the solution?

____

____

____

____ **1**

(4)

[Turn over

DO NOT
WRITE IN
THIS
MARGIN

Marks | KU | PS

19. Esters are compounds used in perfumes.

Esters can be made when an alcohol reacts with a carboxylic acid.

$$
\begin{array}{ccccccc}
\overset{\displaystyle H}{\underset{\displaystyle H}{H-C-O-H}} & + & \overset{\displaystyle O\ H}{\underset{\displaystyle H}{H-O-C-C-H}} & \rightarrow & \overset{\displaystyle H\ \ \ O\ H}{\underset{\displaystyle H\ \ \ \ \ H}{H-C-O-C-C-H}} & + & H\overset{O}{\ \ \ }H
\end{array}
$$

alcohol   carboxylic acid   ester   water

(a) Draw the **full** structural formula for the **ester** produced in the following reaction.

$$
\overset{\displaystyle H\ \ H}{\underset{\displaystyle H\ \ H}{H-C-C-O-H}} + \overset{\displaystyle O\ H\ H}{\underset{\displaystyle H\ H}{H-O-C-C-C-H}} \rightarrow
$$

alcohol   carboxylic acid

1

(b) After some time the esters in perfumes react with water and break down to form the alcohol and carboxylic acid again.

Suggest a name for the type of chemical reaction taking place.

_____

1
(2)

[END OF QUESTION PAPER]

ADDITIONAL SPACE FOR ANSWERS

ADDITIONAL GRAPH PAPER FOR QUESTION 12(*a*)

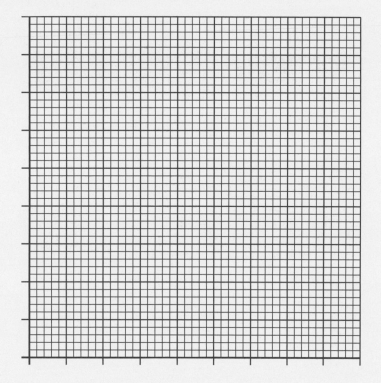

ADDITIONAL SPACE FOR ANSWERS

[BLANK PAGE]

FOR OFFICIAL USE

C

	KU	PS
Total Marks		

0500/402

NATIONAL
QUALIFICATIONS
2007

THURSDAY, 10 MAY
10.50 AM – 12.20 PM

CHEMISTRY
STANDARD GRADE
Credit Level

Fill in these boxes and read what is printed below.

Full name of centre

Town

Forename(s)

Surname

Date of birth

Day	Month	Year		Scottish candidate number		Number of seat

1 All questions should be attempted.

2 Necessary data will be found in the Data Booklet provided for Chemistry at Standard Grade and Intermediate 2.

3 The questions may be answered in any order but all answers are to be written in this answer book, and must be written clearly and legibly in ink.

4 Rough work, if any should be necessary, as well as the fair copy, is to be written in this book.
Rough work should be scored through when the fair copy has been written.

5 Additional space for answers and rough work will be found at the end of the book.

6 The size of the space provided for an answer should not be taken as an indication of how much to write. It is not necessary to use all the space.

7 Before leaving the examination room you must give this book to the invigilator. If you do not, you may lose all the marks for this paper.

SCOTTISH
QUALIFICATIONS
AUTHORITY

PART 1

In Questions 1 to 9 of this part of the paper, an answer is given by circling the appropriate letter (or letters) in the answer grid provided.

In some questions, two letters are required for full marks.

If more than the correct number of answers is given, marks will be deducted.

A total of 20 marks is available in this part of the paper.

SAMPLE QUESTION

A CH_4	B H_2	C CO_2
D CO	E C_2H_5OH	F C

(a) Identify the hydrocarbon.

Ⓐ	B	C
D	E	F

The one correct answer to part (a) is A. This should be circled.

(b) Identify the **two** elements.

A	Ⓑ	C
D	E	Ⓕ

As indicated in this question, there are **two** correct answers to part (b). These are B and F. Both answers are circled.

If, after you have recorded your answer, you decide that you have made an error and wish to make a change, you should cancel the original answer and circle the answer you now consider to be correct. Thus, in part (a), if you want to change an answer A to an answer D, your answer sheet would look like this:

Ⓐ̸	B	C
Ⓓ	E	F

If you want to change back to an answer which has already been scored out, you should enter a tick (✓) in the box of the answer of your choice, thus:

✓Ⓐ̸	B	C
Ⓓ̸	E	F

1.

Testing gases

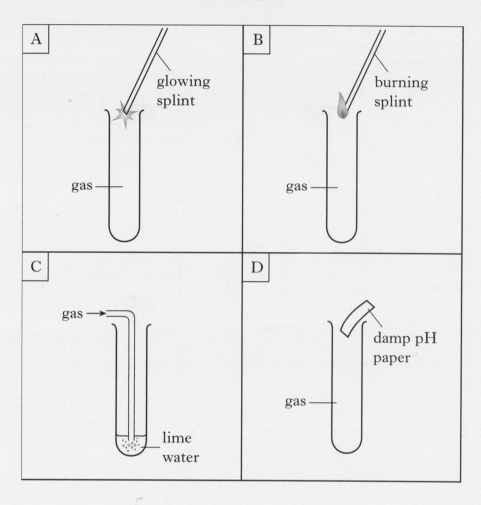

A — glowing splint — gas

B — burning splint — gas

C — gas → lime water

D — damp pH paper — gas

(a) Identify the test for oxygen gas.

A	B
C	D

1

(b) Identify a test for ammonia gas.

A	B
C	D

1

(2)

2

[Turn over

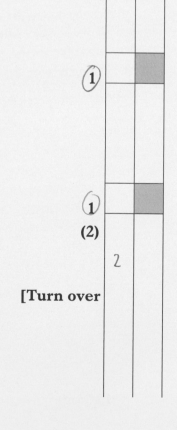

Marks | KU | PS

2. Zinc and magnesium both react with dilute hydrochloric acid.

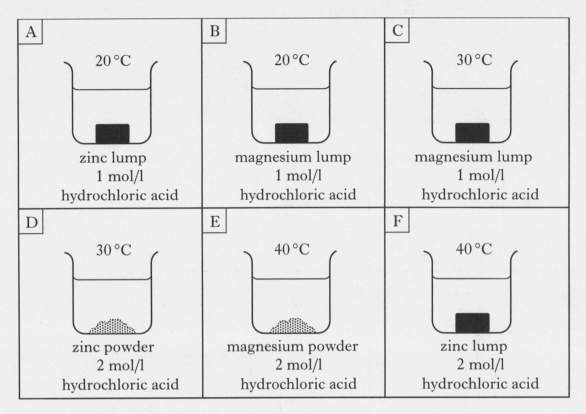

(*a*) Identify the experiment with the **slowest** rate of reaction.

A	B	C
D	E	F

(*b*) Identify the **two** experiments which could be used to investigate the effect of temperature on the rate of reaction.

A	B	C
D	E	F

1

1

(2)

2

Marks | KU | PS

3. Distillation of crude oil produces several fractions.

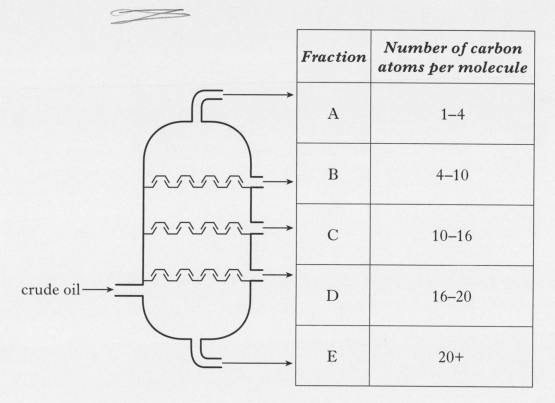

Fraction	Number of carbon atoms per molecule
A	1–4
B	4–10
C	10–16
D	16–20
E	20+

crude oil →

(a) Identify the fraction which is used to tar roads.

A
B
C
(D)
E

1

(b) Identify the fraction with the lowest boiling point.

(A)
B
C
D
E

1

(2)

[Turn over

4. The structural formulae for some hydrocarbons are shown below.

A	B	C
$\begin{array}{c} CH_3 \quad H \\ \mid \qquad \mid \\ C = C \\ \mid \qquad \mid \\ H \qquad H \end{array}$	$\begin{array}{c} H \quad\quad H \\ \diagdown\quad\diagup \\ C \\ \diagup \quad \diagdown \\ H{-}C{-}{-}{-}C{-}H \\ \diagup \qquad \diagdown \\ H \qquad\qquad H \end{array}$	$\begin{array}{c} H \qquad CH_3 \\ \mid \qquad \mid \\ H{-}C{-}{-}C{-}H \\ \mid \qquad \mid \\ CH_3 \quad H \end{array}$
D	E	F
$\begin{array}{c} H \quad H \\ \mid \quad \mid \\ H{-}C{-}C{-}H \\ \mid \quad \mid \\ H{-}C{-}C{-}H \\ \mid \quad \mid \\ H \quad H \end{array}$	$\begin{array}{c} CH_3 \quad H \\ \mid \qquad \mid \\ H{-}C{-}{-}C{-}H \\ \mid \qquad \mid \\ H \qquad H \end{array}$	$\begin{array}{c} CH_3 \quad H \\ \mid \qquad \mid \\ C = C \\ \mid \qquad \mid \\ H \qquad CH_3 \end{array}$

(*a*) Identify the hydrocarbon which could be used to make poly(butene).

A	B	C
D	E	F

(*b*) Identify the **two** hydrocarbons with the general formula C_nH_{2n} which do **not** react quickly with hydrogen.

A	B	C
D	E	F

1

(2)

Marks

5. The table contains information about some substances.

Substance	Melting point/ °C	Boiling point/ °C	Conducts as a solid	Conducts as a liquid
A	1700	2230	no	no
B	605	1305	no	yes
C	−13	77	no	no
D	801	1413	no	yes
E	181	1347	yes	yes
F	−39	357	yes	yes

(a) Identify the substance which exists as covalent molecules.

A
B
C
D
E
F

1

(b) Identify the metal which is liquid at 25 °C.

A
B
C
D
E
F

1

(2)

[Turn over

6. Equations are used to represent chemical reactions.

A	$Sn(s) \longrightarrow Sn^{2+}(aq) + 2e^-$
B	$Cu^{2+}(aq) + 2e^- \longrightarrow Cu(s)$
C	$H^+(aq) + OH^-(aq) \longrightarrow H_2O(\ell)$
D	$2Mg(s) + O_2(g) \longrightarrow 2MgO(s)$
E	$SO_2(g) + H_2O(\ell) \longrightarrow 2H^+(aq) + SO_3^{2-}(aq)$

(a) Identify the equation which represents the formation of acid rain.

A
B
C
D
E

(b) Identify the equation which represents neutralisation.

A
B
C
D
E

(c) Identify the **two** equations in which a substance is oxidised.

A
B
C
D
E

Marks

1

1

2

(4)

Official SQA Past Papers: Credit Chemistry 2007

DO NOT
WRITE IN
THIS
MARGIN

Marks KU PS

7. A student made some statements about the particles found in atoms.

A	It has a positive charge.
B	It has a negative charge.
C	It has a relative mass of almost zero.
D	It has a relative mass of 1.
E	It is found inside the nucleus.
F	It is found outside the nucleus.

Identify the **two** statements which apply to **both** a proton and a neutron.

A
B
C
D
E
F

(2)

8. A student made some statements about the reaction of silver(I) oxide with excess dilute hydrochloric acid.

A	The concentration of hydrogen ions increases.
B	Carbon dioxide gas is produced.
C	An insoluble salt is produced.
D	Hydrogen gas is produced.
E	Water is produced.

Identify the **two** correct statements.

A
B
C
D
E

(2)

Marks KU PS

9. When two different electrodes are joined in a cell, a chemical reaction takes place and a voltage is produced.

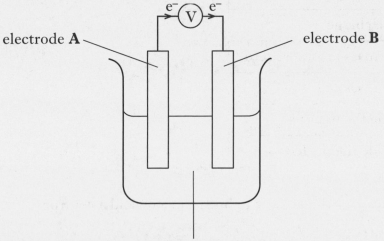

electrode **A**

electrode **B**

sodium chloride solution and ferroxyl indicator

	Electrode A	*Electrode B*
A	magnesium	iron
B	iron	carbon
C	iron	aluminium
D	iron	copper
E	lead	iron

Which **two** pairs of electrodes will produce a flow of electrons in the same direction as shown in the diagram and would produce a blue colour around electrode **A**?

You may wish to use the data booklet to help you.

A
B
C
D
E

 (2)

[Turn over for Part 2 on *Page twelve*

PART 2

A total of 40 marks is available in this part of the paper.

10. A sample of silver was found to contain two isotopes, $^{107}_{47}Ag$ and $^{109}_{47}Ag$.

(a) This sample of silver has an average atomic mass of 108.

What does this indicate about the amount of each isotope in this sample?

_____

_____ —1

(b) Complete the table to show the number of each type of particle in a $^{107}_{47}Ag^+$ ion.

Particle	Number
proton	47
neutron	60
electron	47

2

(c) Silver can be displaced from a solution of silver(I) nitrate.

$$2AgNO_3(aq) + Cu(s) \longrightarrow 2Ag(s) + Cu(NO_3)_2(aq)$$

(i) Balance this equation. 1

(ii) Name a metal which would **not** displace silver from silver(I) nitrate.

You may wish to use the data booklet to help you.

_____ — 1

(5)

Marks KU PS

11. Alkanoic acids are a family of compounds which contain the $-C{\Large\langle}^{O}_{O-H}$ group.

The **full** structural formulae for the first three members are shown.

$H-C{\Large\langle}^{O}_{O-H}$ $H-\overset{\displaystyle H}{\underset{\displaystyle H}{C}}-C{\Large\langle}^{O}_{O-H}$ $H-\overset{\displaystyle H}{\underset{\displaystyle H}{C}}-\overset{\displaystyle H}{\underset{\displaystyle H}{C}}-C{\Large\langle}^{O}_{O-H}$

methanoic   ethanoic   propanoic
acid   acid   acid

(a) Draw the **full** structural formula for the alkanoic acid containing 4 carbon atoms.

⁻1

(b) The table gives information on some alkanoic acids.

Acid	Boiling point/°C
methanoic acid	101
ethanoic acid	118
propanoic acid	141
butanoic acid	164

17
23
23

(i) Using this information, make a general statement linking the boiling point to the number of carbon atoms.

_____

_____

_____   ⁻1

(ii) Predict the boiling point of pentanoic acid.

_____187_____ °C   1

(3)

12. Ammonia is made when nitrogen and hydrogen react together.

The table below shows the percentage yields obtained when nitrogen and hydrogen react at different pressures.

Pressure/atmospheres	Percentage yield of ammonia
25	28
50	40
100	53
200	67
400	80

(a) Draw a line graph of percentage yield against pressure.

Use appropriate scales to fill most of the graph paper.

(Additional graph paper, if required, will be found on page 27.)

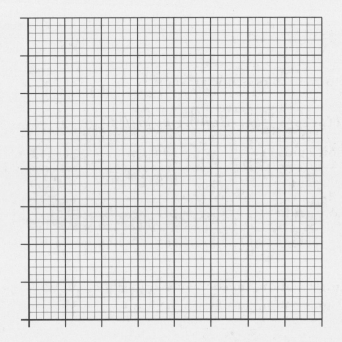

2

(b) Use your graph to estimate the percentage yield of ammonia at 150 atmospheres.

1

12. (continued)

(*c*) Ammonia can be produced in the lab by heating an ammonium compound with soda lime.

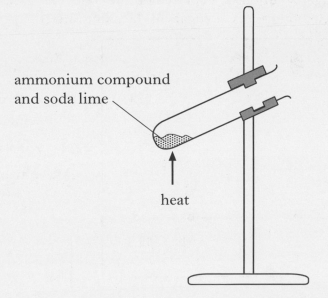

ammonium compound
and soda lime

heat

In order to produce ammonia, what **type** of compound must soda lime be?

_____

1

(4)

[Turn over

13. Starch and sucrose can be hydrolysed to produce simple sugars.

Chromatography is a technique which can be used to identify the sugars produced.

Samples of known sugar solutions are spotted on the base line. The solvent travels up the paper carrying spots of sugars at different rates.

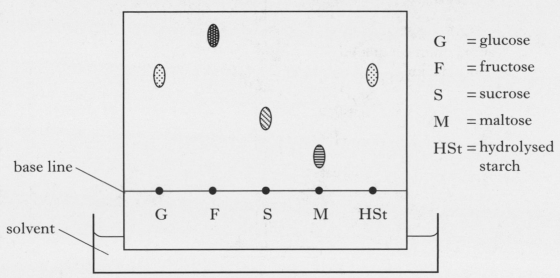

The diagram above shows that **only glucose** is produced when starch is hydrolysed.

(a) The chromatogram below can be used to identify the simple sugars produced when sucrose is hydrolysed.

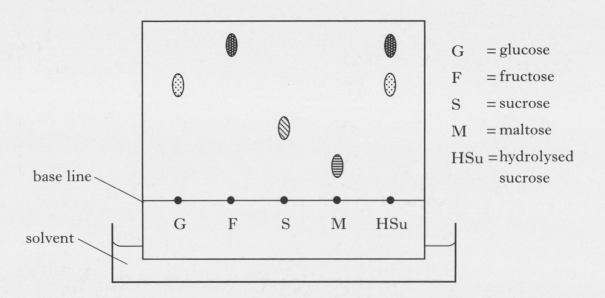

Name the sugars produced when sucrose is hydrolysed.

_____

1

DO NOT
WRITE IN
THIS
MARGIN

Marks | KU | PS

13. (continued)

(*b*) What **type** of substance, present in the digestive system, acts as a catalyst in the hydrolysis of sucrose?

1

(2)

[Turn over

Marks KU PS

14. Cars made from steel can be protected from rusting in a number of ways.

(a) Circle the correct word to complete the sentence below.

Steel does not rust when attached to the $\left\{ \begin{array}{c} \text{negative} \\ \text{positive} \end{array} \right\}$ terminal of a car battery.

1

(b) The steel body of the car can be coated by dipping it in molten zinc.

  (i) What name is given to this process?

  _____

1

  (ii) Explain why the steel does **not** rust even when the zinc coating is scratched.

  _____

  _____

  _____

1

(3)

Marks

| KU | PS |

15. The atoms in a chlorine molecule are held together by a covalent bond. A covalent bond is a shared pair of electrons.

The chlorine molecule can be represented as

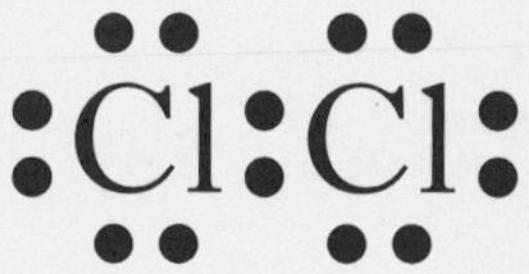

● = electron

(a) Showing **all** outer electrons, draw a similar diagram to represent a molecule of hydrogen chloride, HCl.

1

(b) In forming covalent bonds, why do atoms share electrons?

_____

_____

_____

1

(2)

[Turn over

16. Ethanol is the alcohol found in alcoholic drinks.

It can be produced as shown in the diagram.

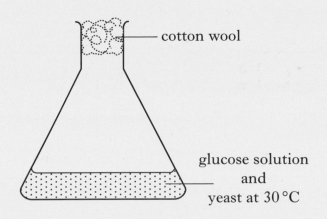

cotton wool

glucose solution and yeast at 30 °C

(a) (i) Name the type of chemical reaction taking place in the flask.

_____ 1

(ii) What would happen to the rate of the reaction if the experiment above was repeated at 50 °C?

_____ 1

(b) In industry, alcohols can be produced from alkenes as shown in the example below.

$$\begin{array}{ccc} & H & H & H \\ & | & | & | \\ H- & C- & C- & C-H \\ & | & | & | \\ & H & H & O\text{-}H \end{array}$$
propan-1-ol

$$\begin{array}{ccc} H & H & H \\ | & | & | \\ H-C-C=C-H \\ | \\ H \end{array} + H\diagup\!\!{}^O\!\!\diagdown H$$
propene   water

$$\begin{array}{ccc} & H & H & H \\ & | & | & | \\ H- & C- & C- & C-H \\ & | & | & | \\ & H & O\text{-}H & H \end{array}$$
propan-2-ol

(i) Name the type of chemical reaction taking place.

_____ 1

DO NOT WRITE IN THIS MARGIN

Marks	KU	PS

16. (*b*) (continued)

(ii) What **term** is used to describe a pair of alcohols like propan-1-ol and propan-2-ol?

______________________________ **1**

(iii) Propan-1-ol and propan-2-ol have different boiling points.

Name the process which could be used to separate a mixture of these alcohols.

______________________________ **1**

(5)

[Turn over

Marks | KU | PS

17. The table contains information on minerals.

Mineral	Formula
cinnabar	HgS
fluorite	CaF_2
gibbsite	$Al(OH)_3$
haematite	Fe_2O_3
zinc blende	ZnS

(a) State the chemical name for zinc blende.

_____

1

(b) Name the salt formed when gibbsite reacts with dilute hydrochloric acid.

_____

1

(c) Calculate the percentage, by mass, of calcium in fluorite (CaF_2).

Show your working clearly.

_____ %

2

(d) Iron metal can be extracted from haematite (Fe_2O_3) by heating with carbon monoxide. Carbon dioxide is also produced.

Write an equation, using **symbols** and **formulae**, for this reaction.

There is no need to balance it.

_____

1

(e) Name a metal which can be extracted from its ore by heat alone.

_____

1

(6)

18. Nylon is a polymer with many uses.

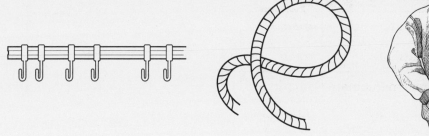

curtain rail   rope   jacket

(a) Nylon is a thermoplastic polymer.

What does thermoplastic mean?

_____

_____   1

(b) Nylon is a polymer made from two different monomers as shown.

$$H-N-(CH_2)_6-N-H \quad H-O-C-(CH_2)_4-C-O-H$$

$$-N-(CH_2)_6-N-C-(CH_2)_4-C- \ + \ H \ \ H$$

nylon

During the polymerisation reaction, water is also produced.

Suggest a name for this **type** of polymerisation.

_____   1
                                                         (2)

[Turn over

Marks | KU | PS

19. Many ionic compounds are coloured.

Compound	Colour
nickel(II) nitrate	green
nickel(II) sulphate	green
potassium permanganate	purple
potassium sulphate	colourless

(a) Using the information in the table, state the colour of the potassium ion.

_____

1

(b) Write the **ionic** formula for nickel(II) nitrate.

_____

1

(c) A student set up the following experiment to investigate the colour of the ions in copper(II) chromate.

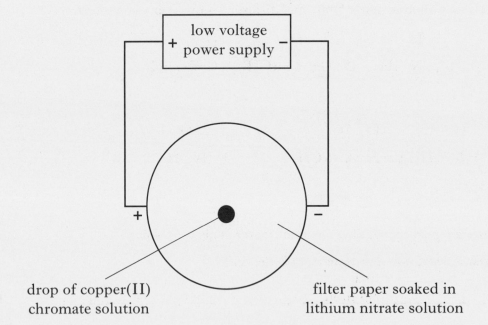

drop of copper(II) chromate solution

filter paper soaked in lithium nitrate solution

The student made the following observation.

Observation
yellow colour moves to the positive electrode
blue colour moves to the negative electrode

DO NOT WRITE IN THIS MARGIN

Marks KU PS

19. (c) (continued)

(i) State the colour of the chromate ion.

______________________________ **1**

(ii) Lithium nitrate solution is used as the electrolyte.

What is the purpose of an electrolyte?

____

____

______________________________ **1**

(iii) Suggest why lithium phosphate can **not** be used as the electrolyte in this experiment.

You may wish to use the data booklet to help you.

____

____

______________________________ **1**

(5)

[Turn over

DO NOT WRITE IN THIS MARGIN

Marks | KU | PS

20. Indigestion is caused by excess acid in the stomach. Indigestion remedies containing calcium carbonate neutralise some of this acid.

Christine carried out an experiment to find the mass of calcium carbonate required to neutralise a dilute hydrochloric acid solution.

She added calcium carbonate until all the acid had been used up.

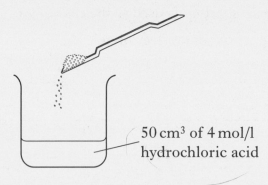

50 cm³ of 4 mol/l hydrochloric acid

(a) Calculate the number of moles of dilute hydrochloric acid used in the experiment.

_____ mol   1

(b) The equation for the reaction is

$$CaCO_3(s) + 2HCl(aq) \longrightarrow CaCl_2(aq) + H_2O(\ell) + CO_2(g)$$

(i) Using your answer from part (a), calculate the number of moles of calcium carbonate required to neutralise the dilute hydrochloric acid.

1 mole   2 moles
0·1 ⟵ 0·2

_____ 0·1 mol   1

(ii) Using your answer from part (b)(i), calculate the **mass** of calcium carbonate (CaCO₃) required to neutralise the acid.

_____ g   1

(3)

[END OF QUESTION PAPER]

ADDITIONAL SPACE FOR ANSWERS

ADDITIONAL GRAPH PAPER FOR QUESTION 12(*a*)

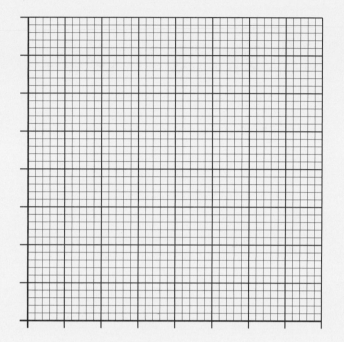

[BLANK PAGE]

[BLANK PAGE]

FOR OFFICIAL USE

C

KU   PS

Total Marks

0500/402

NATIONAL
QUALIFICATIONS
2008

THURSDAY, 1 MAY
10.50 AM – 12.20 PM

CHEMISTRY
STANDARD GRADE
Credit Level

Fill in these boxes and read what is printed below.

Full name of centre

Town

Forename(s)

Surname

Date of birth

Day Month Year   Scottish candidate number   Number of seat

1 All questions should be attempted.

2 Necessary data will be found in the Data Booklet provided for Chemistry at Standard Grade and Intermediate 2.

3 The questions may be answered in any order but all answers are to be written in this answer book, and must be written clearly and legibly in ink.

4 Rough work, if any should be necessary, as well as the fair copy, is to be written in this book.

Rough work should be scored through when the fair copy has been written.

5 Additional space for answers and rough work will be found at the end of the book.

6 The size of the space provided for an answer should not be taken as an indication of how much to write. It is not necessary to use all the space.

7 Before leaving the examination room you must give this book to the invigilator. If you do not, you may lose all the marks for this paper.

PART 1

In Questions 1 to 9 of this part of the paper, an answer is given by circling the appropriate letter (or letters) in the answer grid provided.

In some questions, two letters are required for full marks.

If more than the correct number of answers is given, marks will be deducted.

A total of 20 marks is available in this part of the paper.

SAMPLE QUESTION

A CH_4	B H_2	C CO_2
D CO	E C_2H_5OH	F C

(a) Identify the hydrocarbon.

Ⓐ	B	C
D	E	F

The one correct answer to part (a) is A. This should be circled.

(b) Identify the **two** elements.

A	Ⓑ	C
D	E	Ⓕ

As indicated in this question, there are **two** correct answers to part (b). These are B and F. Both answers are circled.

If, after you have recorded your answer, you decide that you have made an error and wish to make a change, you should cancel the original answer and circle the answer you now consider to be correct. Thus, in part (a), if you want to change an answer A to an answer D, your answer sheet would look like this:

A̸	B	C
Ⓓ	E	F

If you want to change back to an answer which has already been scored out, you should enter a tick (✓) in the box of the answer of your choice, thus:

✓A̸	B	C
D̸	E	F

1. The formulae of some gases are shown in the grid.

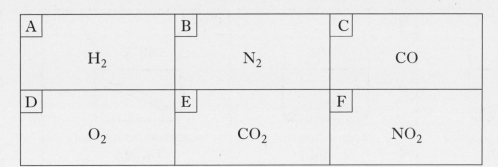

A	B	C
H_2	N_2	CO
D	E	F
O_2	CO_2	NO_2

(a) Identify the toxic gas produced during the burning of plastics.

A	B	Ⓒ
D	E	F

(b) Identify the gas which makes up approximately 80% of air.

A	Ⓑ	C
D	E	F

(c) Identify the gas used up during respiration.

A	B	C
Ⓓ	E	F

[Turn over

2. A student carried out several experiments with metals and acids.

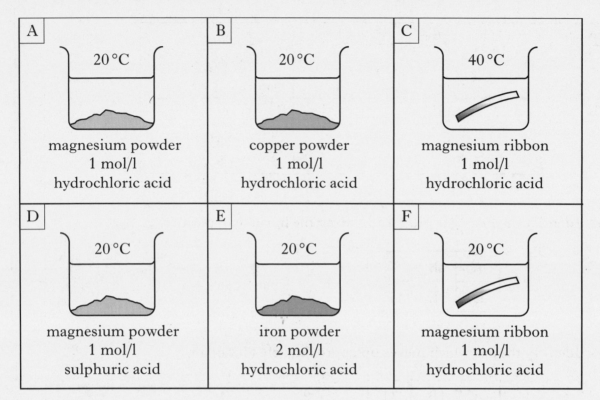

A	B	C
20 °C	20 °C	40 °C
magnesium powder 1 mol/l hydrochloric acid	copper powder 1 mol/l hydrochloric acid	magnesium ribbon 1 mol/l hydrochloric acid

D	E	F
20 °C	20 °C	20 °C
magnesium powder 1 mol/l sulphuric acid	iron powder 2 mol/l hydrochloric acid	magnesium ribbon 1 mol/l hydrochloric acid

(*a*) Identify the **two** experiments which could be compared to show the effect of particle size on reaction rate.

Ⓐ	B	C
D	E	Ⓕ

1

(*b*) Identify the experiment in which **no** reaction would take place.

A	Ⓑ	C
D	E	F

(2)

Marks | KU | PS

3. The grid shows the structural formulae of some hydrocarbons.

A
$$H-\overset{\overset{\displaystyle H}{|}}{\underset{\underset{\displaystyle H}{|}}{C}}-\overset{\overset{\displaystyle H}{|}}{\underset{\underset{\displaystyle H}{|}}{C}}-H$$

B
(cyclopropane structure with three C atoms each bearing H atoms)

C
$$\overset{\overset{\displaystyle H}{|}}{\underset{\underset{\displaystyle H}{|}}{C}}=\overset{\overset{\displaystyle H}{|}}{\underset{\underset{\displaystyle H}{|}}{C}}$$

D
$$\overset{\overset{\displaystyle CH_3}{|}}{\underset{\underset{\displaystyle H}{|}}{C}}=\overset{\overset{\displaystyle H}{|}}{\underset{\underset{\displaystyle H}{|}}{C}}$$

E
$$H-\overset{\overset{\displaystyle H}{|}}{\underset{\underset{\displaystyle H}{|}}{C}}-\overset{\overset{\displaystyle CH_3}{|}}{\underset{\underset{\displaystyle CH_3}{|}}{C}}-\overset{\overset{\displaystyle H}{|}}{\underset{\underset{\displaystyle H}{|}}{C}}-H$$

F
$$H-\overset{\overset{\displaystyle H}{|}}{\underset{}{C}}-\overset{\overset{\displaystyle H}{|}}{\underset{}{C}}-H$$
$$H-\overset{}{\underset{\underset{\displaystyle H}{|}}{C}}-\overset{}{\underset{\underset{\displaystyle H}{|}}{C}}-H$$

(a) Identify the **two** hydrocarbons which can polymerise.

A	B	Ⓒ
Ⓓ	E	F

1

(b) Identify the **two** hydrocarbons with the general formula C_nH_{2n} which do **not** decolourise bromine solution quickly.

A	Ⓑ	C
D	E	Ⓕ

1

(2)

[Turn over

Marks KU PS

4. The grid shows the names of some oxides.

A	B	C
silicon dioxide	carbon dioxide	sodium oxide ✗
D	E	F
iron oxide ✗	sulphur dioxide	copper oxide ✗

(a) Identify the **two** oxides which contain transition metals.

You may wish to use the data booklet to help you.

A	B	C
Ⓓ	E	Ⓕ

1

(b) Identify the oxide which reacts with water in the atmosphere to produce acid rain.

A	B	C
D	Ⓔ	F

1

(c) Identify the oxide which, when added to water, produces a solution with a greater concentration of hydroxide ions (OH^-) than hydrogen ions (H^+).

A	B	Ⓒ
D	E	F

①

(3)

5. There are different types of chemical reactions.

A	redox
B	precipitation
C	combustion
D	neutralisation
E	displacement

(a) Identify the type of chemical reaction taking place when dilute hydrochloric acid reacts with a carbonate.

A
B
C
(D)
E

1

(b) Identify the **two** types of chemical reaction represented by the following equation.

$$2Zn(s) + O_2(g) \longrightarrow 2ZnO(s)$$

(A)
B
(C)
D
E

2

(3)

[Turn over

DO NOT WRITE IN THIS MARGIN

Marks | KU | PS

6. Lemonade can be made by dissolving sugar, lemon and carbon dioxide in water.

A	sugar
B	lemon
C	carbon dioxide
D	water

Identify the solvent used to make lemonade.

A
B
C
D

(1)

7. The grid contains the names of some carbohydrates.

A	fructose
B	glucose
C	maltose
D	sucrose
E	starch

(*a*) Galactose is a monosaccharide found in dairy products.

Identify the **two** isomers of galactose.

½

1

(*b*) Identify the carbohydrate which is a condensation polymer.

1

(2)

[Turn over

8. A student made some statements about acids.

A	Acid rain will have no effect on iron structures. ✗
B	A base is a substance which can neutralise an alkali.
C	Treatment of acid indigestion is an example of neutralisation.
D	In a neutralisation reaction the pH of the acid will fall towards 7.
E	When dilute nitric acid reacts with potassium hydroxide solution, the salt potassium nitrate is produced. ✗

Identify the **two** correct statements.

A
B
C
Ⓓ
Ⓔ

(2)

9. Coffee manufacturers have produced a self-heating can of coffee.

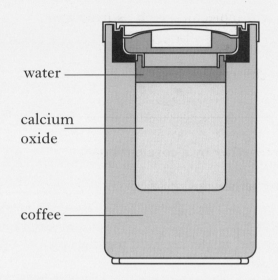

water —

calcium
oxide

coffee —

In the centre of the can calcium oxide reacts with water, releasing heat energy.

The equation for the reaction is:

$$CaO(s) + H_2O(\ell) \longrightarrow Ca(OH)_2(aq)$$

A	Calcium oxide is insoluble.
B	The reaction is exothermic.
C	The reaction produces an acidic solution.
D	The temperature of the coffee goes down.
E	0·1 moles of calcium oxide reacts with water producing 0·1 moles of calcium hydroxide.

Identify the **two** correct statements.

A
B
C
D
E

$18\frac{1}{2}$

$\overline{20}$

$\frac{8\frac{1}{2} \text{ PS}}{9}$

$\frac{10 \text{ kU}}{10}$

(2)

DO NOT
WRITE IN
THIS
MARGIN

Marks | KU | PS

PART 2

A total of 40 marks is available in this part of the paper.

10. Hydrogen reacts with other elements to form molecules such as hydrogen fluoride and hydrogen chloride.

   (*a*) Name the family to which fluorine and chlorine belong.

   Halogens

   1

   (*b*) The atoms in these molecules are held together by a covalent bond.

   Circle the correct words to complete the sentence.

   A covalent bond forms when two 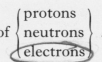 nuclei are held together by

   their common attraction for a shared pair of $\begin{Bmatrix} protons \\ neutrons \\ electrons \end{Bmatrix}$.

   1

   (*c*) The table gives information about some molecules.

Molecule H–X	Size of X/pm	Energy to break bond kJ/mol
H–F	71	569
H–Cl	99	428
H–Br	114	362
H–I	133	295

   Describe how the size of element **X** affects the energy needed to break the bond in the molecule.

   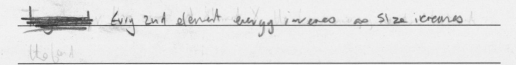 Evry 2nd element energy invenes as Size icreanes

   Halord

   1

   (3)

DO NOT WRITE IN THIS MARGIN

Marks | KU | PS

11. Crude oil can be transported to a refinery through a steel pipeline.

(a) If the pipeline is not protected the iron will rust.

Name the **ion** formed from water and oxygen, when they accept electrons during rusting.

Hydroxide ion

1

(b) Some parts of the pipeline are under the sea.

What effect would seawater have on the rate of rusting?

It would speed it up

1

(c) Magnesium can be attached to the steel pipeline to prevent rusting.

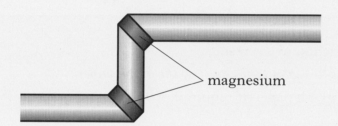

magnesium

What name is given to the **type** of protection provided by the magnesium?

Sacrificial

1

(3)

[Turn over

Marks KU PS

12. Airbags in cars are designed to prevent injuries in car crashes.

They contain sodium azide (NaN_3) which produces nitrogen gas on impact.

The nitrogen inflates the airbag very quickly.

(*a*) The table gives information on the volume of nitrogen gas produced.

Time/microseconds	Volume of nitrogen gas produced/litres
0	0
5	46
10	64
15	74
20	82
25	88
30	88

(i) Draw a line graph of the results.

Use appropriate scales to fill most of the graph paper.

(Additional graph paper, if required, will be found on page 28.)

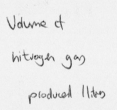

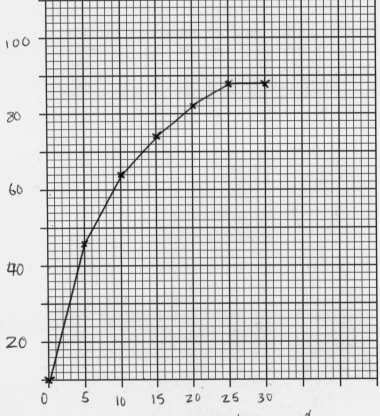

2

(ii) Using your graph, predict the time taken to produce 70 litres of nitrogen gas.

_____ 13 _____ microseconds

1

DO NOT
WRITE IN
THIS
MARGIN

Marks | KU | PS

12. (continued)

(b) The equation for the production of nitrogen gas is:

$$2NaN_3(s) \longrightarrow 3N_2(g) + 2Na(s).$$

Balance the equation above.

1

(c) Nitrogen is a non-toxic gas.

Suggest another property of nitrogen which makes it a suitable gas for use in airbags.

_____not flammable_____

_____

1

(5)

[Turn over

Official SQA Past Papers: Credit Chemistry 2008

DO NOT
WRITE IN
THIS
MARGIN

Marks | KU | PS

13. Copper chloride solution can be broken up into its elements by passing electricity through it.

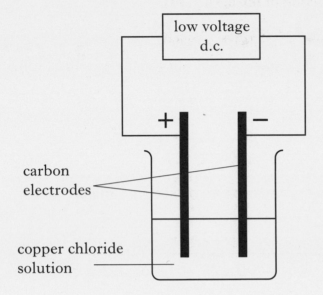

low voltage
d.c.

+   −

carbon
electrodes

copper chloride
solution

(a) Carbon is unreactive and insoluble in water.

Give another reason why it is suitable for use as an electrode.

Graphite conducts electricity   1

(b) Chlorine gas is released at the positive electrode.

Write an ion-electron equation for the formation of chlorine.
You may wish to use the data booklet to help you.

$2Cl^-(aq) \longrightarrow Cl_2(g) + 2e^-$ 1

(c) Why do ionic compounds, like copper chloride, conduct electricity when in solution?

Ions free to move   1
                                                                    (3)

Marks | KU | PS

14. A fizzy drink "Fizz Alive" contains a carbohydrate.

(a) Name all the elements found in a carbohydrate.

_____Carbon hydrogen + oxygen_____

1

(b) A student carried out an investigation to find out which carbohydrate was present in "Fizz Alive".

Test 1   **Test 2**

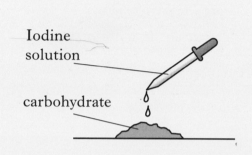

Iodine
solution

carbohydrate

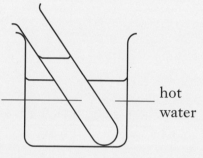

carbohydrate
and Benedict's
solution

hot
water

The results are shown in the table.

Test	Result
Iodine solution	stays brown
Benedict's solution	stays blue

Name the carbohydrate present in "Fizz Alive".

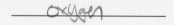
_____Oxygen_____

1

(c) A 330 cm^3 can of "Fizz Alive" has a carbohydrate concentration of 0·01 mol/l.

Calculate the number of <u>moles</u> of carbohydrate in the can of "Fizz Alive".

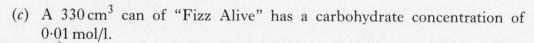

$$n = C \times V$$

$$= 0.01 \quad 0.33$$

_____ mol

1

(3)

[Turn over

Marks | KU | PS

15. The diagram represents the structure of an atom.

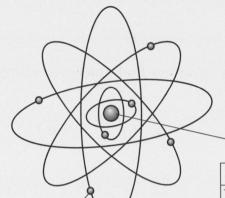

THE NUCLEUS	
Name of Particle	Relative mass
PROTON	(i)
NEUTRON	1

OUTSIDE THE NUCLEUS	
Name of Particle	Relative mass
(ii)	0

(*a*) Fill in the missing information for:

(i) electrons

(ii) Election

1

Marks

KU | PS

15. **(continued)**

(*b*) The element uranium has unstable atoms.

These atoms give out radiation and a new element is formed.

$$^{238}_{92}U \longrightarrow \ ^{234}_{90}Th \ + \ ^{4}_{2}\alpha$$

radiation

(i) Complete the table to show the number of each type of particle in

$^{234}_{90}$Th.

Particle	Number
proton	90
neutron	144

(ii) Radon is another element which gives out radiation.

$$^{222}_{86}Rn \longrightarrow X \ + \ ^{4}_{2}\alpha$$

radiation

State the **atomic number** of element **X**.

_____

1

(3)

[Turn over

DO NOT
WRITE IN
THIS
MARGIN

Marks | KU | PS

16. Anglesite is an ore containing lead(II) sulphate, $PbSO_4$.

(a) Calculate the percentage by mass of lead in anglesite.

Pb SO4

207 + 32 + 64

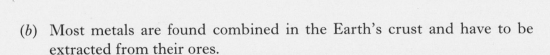

% of element = $\frac{\text{mass of element}}{\text{formula mass}} \times 100$

$= \frac{207}{303} \times 100 = 68.3\%$

68.3 % ⓶

(b) Most metals are found combined in the Earth's crust and have to be extracted from their ores.

Place the following metals in the correct space in the table.

lead   **aluminium**

You may wish to use the data booklet to help you.

Metal	Method of extraction
Aluminium	electrolysis of molten compound
lead	using heat and carbon

①

(c) Metal **X** can be extracted from its ore by heat alone.

What does this indicate about the reactivity of **X** compared to both lead and aluminium?

It's less reactive

①

(d) When a metal is extracted from its ore, metal ions are changed to metal atoms.

Name this **type** of chemical reaction.

Reduction

①
(5)

Marks | KU | PS

17. A student added strips of magnesium to solutions of other metals.

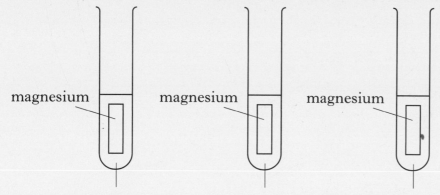

magnesium   magnesium   magnesium

magnesium nitrate   zinc nitrate   copper nitrate
solution   solution   solution

The results are shown in the table.

Solution / Metal	magnesium nitrate	zinc nitrate	copper nitrate
magnesium	(i) ho	(ii) yo	reaction occurred

(a) In the table, fill in the missing information at (i) and (ii) to show whether or not a chemical reaction has occurred.

You may wish to use the data booklet to help you.

1

(b) The equation for the reaction between magnesium and copper nitrate is:

$$Mg(s) + Cu^{2+}(aq) + 2NO_3^-(aq) \longrightarrow Mg^{2+}(aq) + 2NO_3^-(aq) + Cu(s).$$

(i) Circle the spectator ion in the above equation.

1

(ii) What technique could be used to remove copper from the mixture?

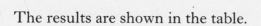
Filtration

1

(3)

[Turn over

18. Nitrogen is essential for healthy plant growth.

Nitrogen from the atmosphere can be fixed in a number of ways.

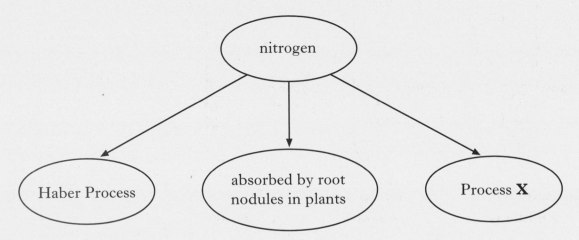

(a) **X** is a natural process which takes place in the atmosphere, producing nitrogen dioxide gas.

What provides the energy for this process?

 Lightning

1

(b) What is present in the root nodules of some plants which convert nitrogen from the atmosphere into nitrogen compounds?

 glucose

1

(c) The Haber Process is the industrial method of converting nitrogen into a nitrogen compound.

Name the nitrogen compound produced.

Ammonia

1

Marks | KU | PS

18. (continued)

(d) The nitrogen compound produced in the Haber Process dissolves in water.

The graph shows the solubility of the nitrogen compound at different temperatures.

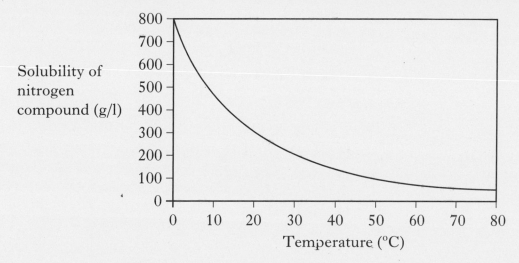

Solubility of nitrogen compound (g/l)

Temperature (°C)

Write a general statement describing the effect of temperature on the solubility of the nitrogen compound.

The higher the temp, the less soluble the ammonia is

①
(4)

[Turn over

Marks | KU | PS

19. The octane number indicates how efficiently a fuel burns.

Alkane	Molecular Formula	Full Structural Formula	Octane Number
2-methylbutane	C_5H_{12}	(see structure)	93
2-methylpentane	C_6H_{14}	(see structure)	71
2-methylhexane	C_7H_{16}	(handwritten structure)	47
2-methylheptane	C_8H_{18}	(see structure)	2
2-methyloctane	C_9H_{20}	(see structure)	2

(*a*) Draw the **full** structural formula for 2-methylhexane.

1

Marks | KU | PS

19. (continued)

(b) 2-methylpentane and hexane have the same molecular formula (C_6H_{14}), but different structural formulae.

What term is used to describe this pair of alkanes?

_____Isomers_____

1

(c) Using information in the table, predict the octane number for 2-methylheptane.

_____21·5_____

1

(3)

[Turn over

DO NOT
WRITE IN
THIS
MARGIN

Marks KU PS

20. Molten iron is used to join steel railway lines together.

Molten iron is produced when aluminium reacts with iron oxide.

The equation for the reaction is:

$2Al + Fe_2O_3 \longrightarrow 2Fe + Al_2O_3$

(a) Calculate the mass of iron produced from 40 grams of iron oxide.

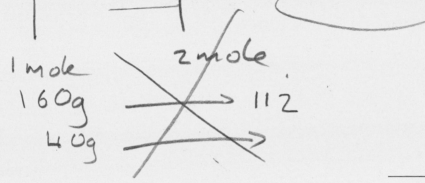

1 mole
160 g
40 g

2 mole
112

_____ g   2

(b) The formula for iron oxide is Fe_2O_3.

What is the charge on this iron ion?

_____ + _____   1

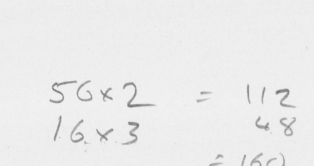

56 × 2 = 112
16 × 3   48
= 160

Marks | KU | PS

20. (continued)

(c)   Iron can also be produced from iron ore, Fe_2O_3, in a blast furnace.

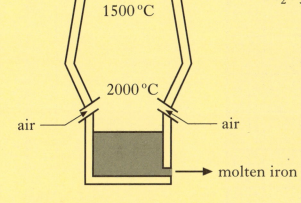

iron ore, carbon and limestone

1000 °C

1500 °C

2000 °C

air

air

molten iron

The main reactions taking place are:

$$C(s) + O_2(g) \longrightarrow CO_2(g)$$

$$CO_2(g) + C(s) \longrightarrow 2CO(g)$$

$$Fe_2O_3(s) + 3CO(g) \longrightarrow 2Fe(\ell) + 3CO_2(g)$$

(i)   When air is blown into the furnace the temperature rises.

Suggest another reason why **air** is blown into the furnace.

Air contains oxygen which is needed in the reaction

1

(ii)   Explain why the temperature at the bottom of the blast furnace should **not** drop below 1535 °C.

You may wish to use the data booklet to help you.

As this is the temp iron melts at

1

(5)

PS
17
20

KU
17
20

[END OF QUESTION PAPER]

ADDITIONAL SPACE FOR ANSWERS

ADDITIONAL GRAPH PAPER FOR QUESTION 12(*a*)(i)

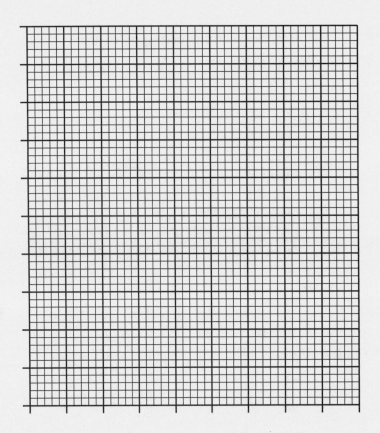

Part 1

PS
$\dfrac{8\frac{1}{2}}{9}$

KU
$\dfrac{10}{11}$

PS Part 2 KU
$\dfrac{17}{20}$

$\dfrac{17}{20}$

$\dfrac{52\frac{1}{2}}{60}$

ADDITIONAL SPACE FOR ANSWERS

[BLANK PAGE]

[BLANK PAGE]

Acknowledgements

Leckie and Leckie is grateful to the copyright holders, as credited, for permission to use their material.

The following companies have very generously given permission to reproduce their copyright material free of charge: BMW(UK) Ltd for an illustration of a BMW car (2007 Credit paper p 18).